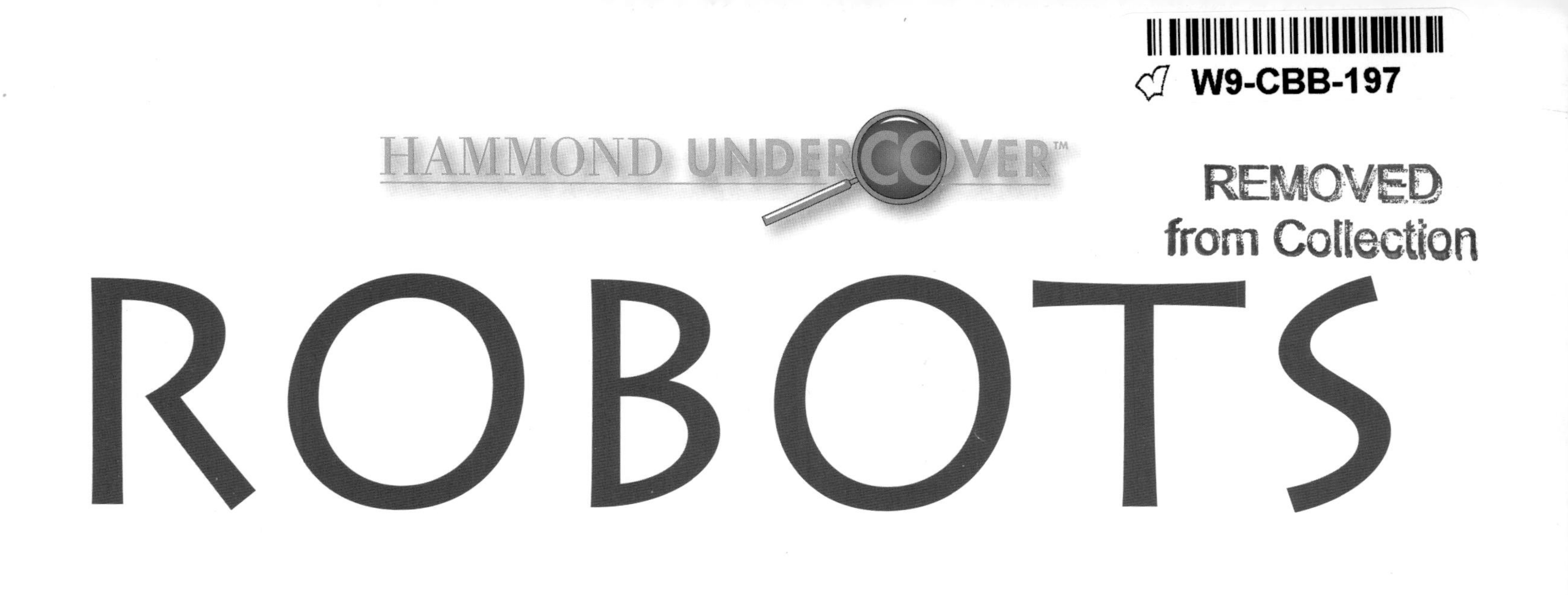

ROBOTS

L HAMMOND World Atlas
Part of the Langenscheidt Publishing Group

Published in the United States and its territories and Canada by
HAMMOND WORLD ATLAS CORPORATION
Part of the Langenscheidt Publishing Group
36-36 33rd Street, Long Island City, NY 11106
EXECUTIVE EDITOR Nel Yomtov
EDITOR Kevin Somers

Produced for Hammond World Atlas Corporation by

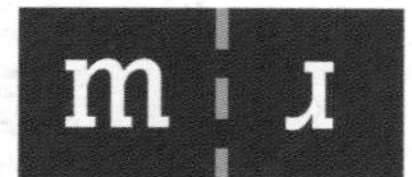

MOSELEY ROAD INC.
129 MAIN STREET
IRVINGTON, NY 10533
WWW.MOSELEYROAD.COM

MOSELEY ROAD INC.
PUBLISHER Sean Moore
ART DIRECTORS Brian MacMullen, Gus Yoo
EDITORIAL DIRECTOR Lisa Purcell

EDITOR Ward Calhoun
PHOTO RESEARCHER Ben DeWalt
PRODUCTION DESIGNER Amy Pierce
CARTOGRAPHY Neil Dvorak
CONTRIBUTING WRITER Frank Vizard
EDITORIAL ASSISTANTS Rachael Lanicci, Natalie Rivera

Printed and bound in Canada

ISBN-13: 978-0841-611368

HAMMOND UNDERCOVER™

ROBOTS

RICK ALLEN LEIDER

Contents

What Is a Robot?

A robot arm, such as this one, was built to weld automotive components. The same kind of arms are shown below, busy at work in an auto factory. Most of the robots in use today build things or perform other factory tasks.

It would probably be impossible to come up with a definition of "robot" that would satisfy everyone. Even robot engineers—the people who build them—use varying definitions of the term. We can say, though, that a robot is a machine that senses, recognizes, and interacts with its environment. That means it can move on its own and may even seem to have a mind of its own!

How a Robot "Thinks"

But even though a robot may move and "think," it is only as smart as its programming. Someone had to design them to do the things they do. Robot engineers design some robots to mimic human behavior, like greeting people or playing a musical instrument. Others program robots to carry out useful tasks, like performing surgery or welding mechanical components. Factory robots can execute complex tasks more quickly and accurately than human workers, and they don't take lunch hours, bathroom breaks, or vacations. About 90 percent of robots work in factories, and half of them build automobiles or work in metal foundries.

EVERYDAY ROBOTS

EVERY DAY you use at least a dozen things created by robots in factories. When you drive in a car or ride a bus, you're traveling in something built mostly by robots.

What Robots Can Do

Still, robots are breaking into many other fields. Robots can do jobs that are very dangerous to humans or ones that humans just find too unpleasant. For example, robots defuse bombs, clean up nuclear waste, and let medical students poke them with needles to practice taking "blood." Someday soon, we'll be seeing robot babysitters, messengers, teachers, and more.

Some robots are built to look like humans, but others are not. But whether they look like mechanical men or merely metal arms, robots are pretty amazing!

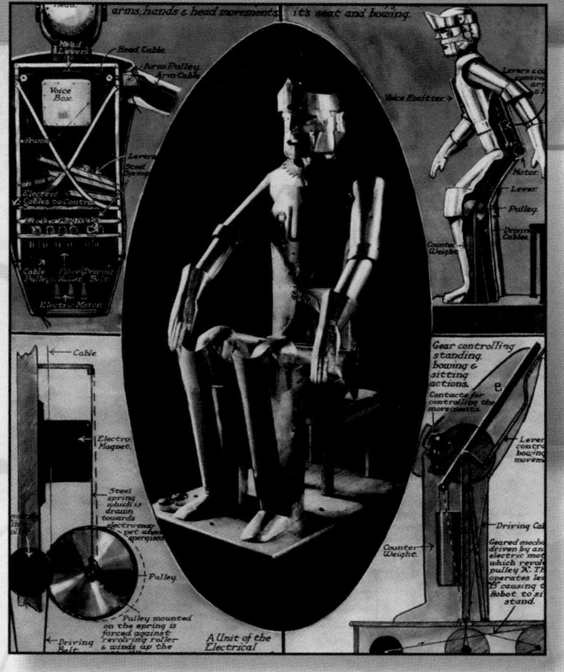

Plans for Rossum's Universal Robot, known as an R.U.R.

ROBOT DATA

THE WORD *ROBOT* comes from the Czechoslovakian *robotnik*, meaning "workman." Writer Karel Capek invented the word for his 1921 play *R.U.R.*, which featured robots that revolt against their creators.

DID YOU KNOW?

When you see most robots, you can tell they're not human, but that's not necessarily the case with an android. An android can look and behave exactly like a real person. Wall-E is a robot and looks like a machine, while TV's *Star Trek: The Next Generation*'s character Lt. Commander Data is an android. The Terminator is also an android.

Lt. Commander Data of *Star Trek* is an android.

USEFUL TERMS TO KNOW

WHEN TALKING about robots, many terms come up that may seem confusing. For instance, what is an android? Is it different from a robot? What about a 'droid? Is that something different? Here's a list to help you figure them out.

Android. An android is a robot that is designed to look and act like a human. So far, they mostly exist only in science fiction, but many robot engineers are working on producing real ones. The shortened form *'droid* or *droid* comes from the Star Wars movies. Droids in that series are very intelligent robots, but don't really look human. They have a lot of uses—from flying spacecraft or translating different languages. Droids only exist in fiction.

Artificial Intelligence. Artificial Intelligence, or "AI" for short, is the science and engineering of making intelligent machines, or ones that can think on their own.

Cyborg. Cyborgs are organisms that have both artificial and natural systems. In fiction, cyborgs combine both systems almost seamlessly, such as the Bionic Woman and *Star Trek* Borgs of TV fame. In real life, "cyborg" is used to refer to a man or woman with bionic, or robotic, implants.

Bionics. In medicine, "bionics" means the replacement or enhancement of organs or other body parts by mechanical versions, such as a prosthetic hand or artificial heart.

Robot Roots

We may now take the existence of robots for granted, but a lot of thought and work went into their development. What would (and could) they really do? How would they be powered? What would they look like? These questions were asked centuries before humans had the know-how to answer any of them.

Consisting of such things as bodies, brains, and sensory systems, the basic elements that robots would need to function were already present in humans. It was Leonardo da Vinci who recognized this and created a blueprint for a mechanical man.

By the early 1900s, there were lots of plans for robots, but many never left the drawing board. Others, like Elektro, a moving robot that proved a big hit at the 1939 New York World's Fair, were built. Toward the end of the twentieth century, those early dreams were realized, and all kinds of robots were rolling off production lines and into modern society.

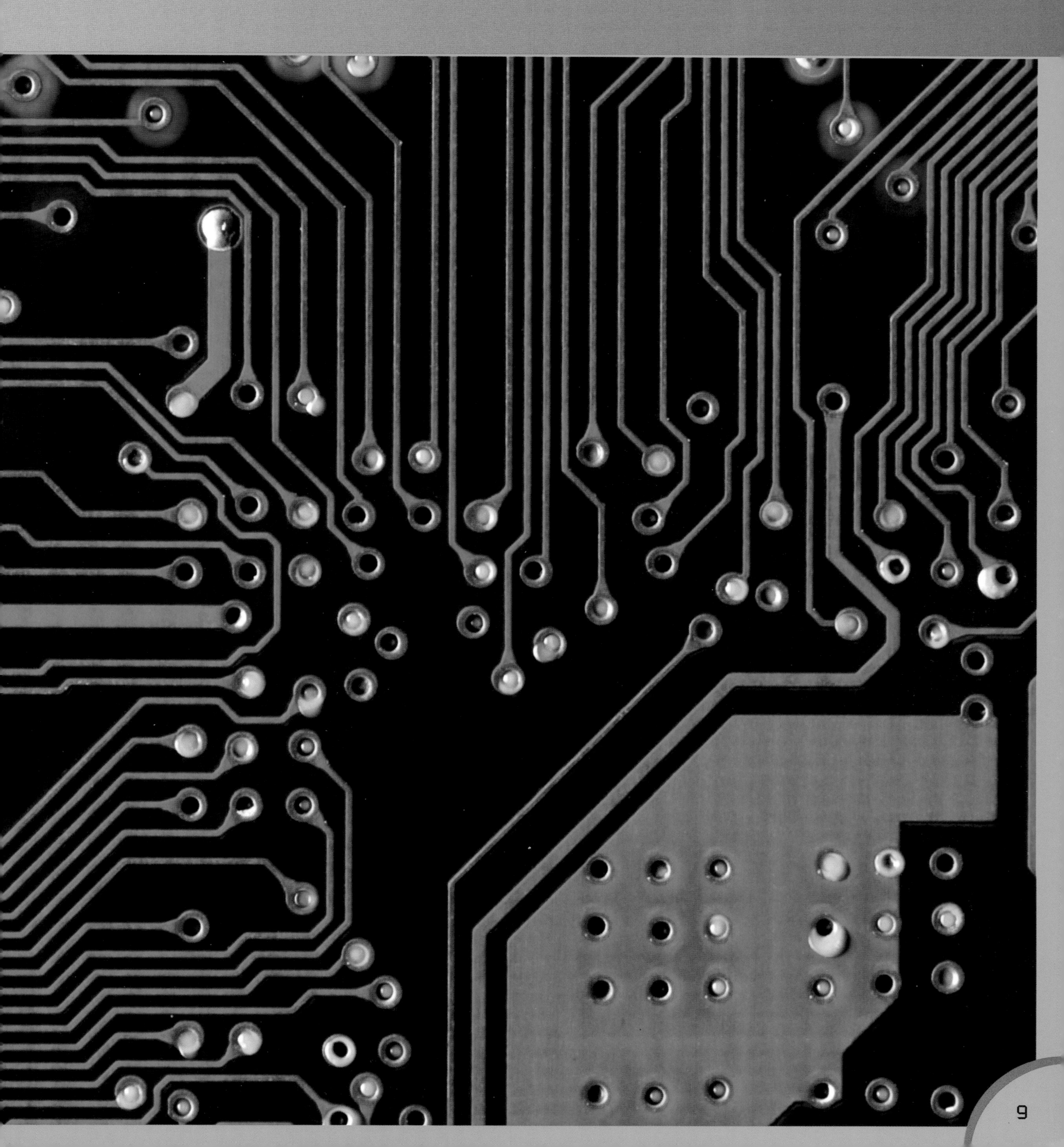

How Robots Work

A robot can be any shape imaginable to fit the requirements of a specific job, but designers keep coming back to the human form for a number or reasons. A robot in human form can use existing tools already made for humans and learn to navigate stairs and corridors already in place. And the human form makes it easier for us to interact with the robot on an emotional level. We're more accepting of something that looks like ourselves.

Robot Basics

Robot designers also may like the human form because humans and robots are very similar. The basic building blocks for a robot are a body or skeleton; some type of muscle system; sensors that supply vision, sound, or feeling; an electrical power supply or rechargeable battery that works like the heart to provide energy; and a computer that works like a brain.

This robot created at the University of Tokyo does his best to look human.

Even the smallest robots have intricate wiring and workings, as illustrated by this x-ray image.

Can Robots Evolve?

ADD A SENSOR or a limb to a robot and you have to reprogram the robot's computer brain to handle the changes. The trouble is that this task is time-consuming and expensive. But what if a robot's software brain could evolve on its own as the robot grew in size and complexity? Scientists at Robert Gordon University at Aberdeen in the United Kingdom have developed software that allows the robot's brain to mimic a human's learning process. For instance, a complex math formula allows the robot to assign a portion of its processors to learn how to cope with the addition of new legs. Over a period of time, the robot essentially teaches itself how to walk much like a human baby.

While that sounds simple, getting a robot to do things we take for granted is more complicated than it looks. For a robot to move its legs, for example, some type of motorized piston needs to be attached to the leg for it to work like a muscle. The movements of the pistons also need to be coordinated with a gyroscope so the robot can keep its balance.

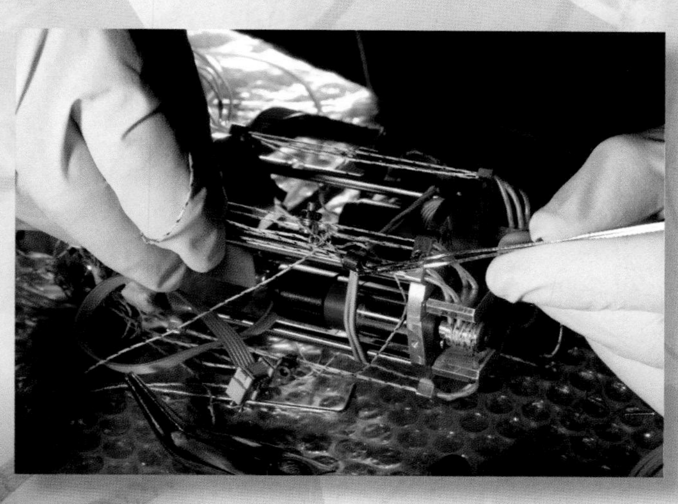

Lots of work goes into the construction of a robot, where even the smallest parts need the attention of skilled technicians.

RISE OF THE SWARM

SEND 200 ROBOTS to find something and they'll do the job quicker than just one. There are two problems, however: how do these robots "talk" to each other and how do they stop from bumping into each other? Scientists are taking a cue from bees to develop robot swarms. This "beeware" allows robots to keep their distance from one another while staying close enough to communicate. One method detects changes in light to judge how close another robot is while a second uses reflecting sound waves. Chirping sounds might also be used to establish location and communication. Another technique uses a smart "queen bee" robot to deploy and monitor a squad of dumber robots while also providing a radio link to human controllers.

Though robot swarms have yet to be perfected, that hasn't stopped computer artists from imagining what they might look like.

This Makes Sense

Sensors that tell the robot about the world are very important. A robot's eyes work much like a digital camera, but the tricky part is teaching the robot context. Is the poster of a lion on the wall a real lion or just a picture of one? Likewise, a robot's ear can work like a microphone, but being able to tell the difference between commands and background noise is the challenge.

Another sense we take for granted is touch. Robots, however, require special fingertip sensors that emit a tiny electrical field. This allows the robot to judge the size and shape of an object and to adjust their fingers accordingly so it can be gripped.

Though scientists have not figured a way to make LEGO blocks move on their own, they have succeeded in making robots, such as these vehicles, out of LEGO.

A battery, located down near the wrist, is the power source for this robot hand.

SHAPE CHANGERS

IMAGINE A ROBOT that could change shape on demand. Such a robot wouldn't actually be just one machine but a collection of mini-robots about a millimeter in size. Researchers at Carnegie Mellon University are designing tiny round robots with a ring of electromagnets around their edges. By switching the electromagnets on and off, the mini-robots can attach themselves and roll around each other to form shapes. Think of it as getting LEGO blocks to form things on their own. Scientists ultimately hope to develop "smart sand" that can arrange itself into structures and communicate among its "grains" to respond to changes in its environment.

Taste, of course, is less important because robots don't eat food to supply their bodies with energy. Still, a robot could be built with taste sensors if it were required for a specific job such as identifying toxins in food.

Brain Power

Electronic circuits link all these functions to a programmable computer that acts as the robot's brain. The microprocessors send the commands that allow a robot to perform a function. A harder task is getting a robot's brain to remember where it's been before. Scientists are devising software programs that allow the robot to assign as many as 1,000 words to a visual image of a location so that it can remember it's been there before. This is what memories are made of, if you're a robot.

DID YOU KNOW?

Courses on how a robot works may soon be taught by instructors like Saya, a robot teacher already taking over routine tasks like taking attendance in Japanese primary schools. Developed by Professor Hiroshi Kobayashi at the University of Tokyo, Saya can mimic a wide range of human emotions—including anger at uncooperative students—thanks to 18 small motors hidden beneath the latex skin covering her face. The Japanese government hopes to use robots to fill jobs left vacant by an anticipated labor shortage due to an aging population.

Initially introduced as a receptionist robot, Saya is now taking her act to Japanese schools.

Early Ideas for Robotics

Throughout history, people have pondered the concept of artificial life, but it was around 1495 when Renaissance genius Leonardo da Vinci put his ideas on paper. The painter of the *Mona Lisa* and *The Last Supper* had become an expert on human anatomy by dissecting real, dead bodies. He put this knowledge to use by writing many pages of notes and sketching numerous drawings for a mechanical man.

He designed his humanoid robot to be constructed in a suit of armor. It worked by a system of gears and pulleys linked with cables and clockworks. Unfortunately, da Vinci never actually built his mechanical man.

Robot Drama

In 1920, hundreds of years after da Vinci's work, Czech writer Karel Capek wrote a play, *R.U.R.* (Rossum's Universal Robots), about humanoid robots that turn against their creators. This play led people to think that robots were supposed to do the difficult jobs that humans couldn't and that these same robots could become very dangerous.

Decades later, writer-philosopher Isaac Asimov had a very different idea. In his

One of the robots from *R.U.R.* by Karel Capek. Capek came up with the term *robot* for his mechanical creatures.

GUESS WHAT?

FOLLOWING CAPEK'S *R.U.R.*, German director Fritz Lang featured a 6-foot robot in his classic 1929 film *Metropolis*, which also contained the first movie android. Lang's film showed the evils and abuses of the Industrial Revolution and science of the 1900s and warned of its possible dangers. When scientist Rotwang transforms his robot into a look-alike of the human heroine Maria, cinema had its first android.

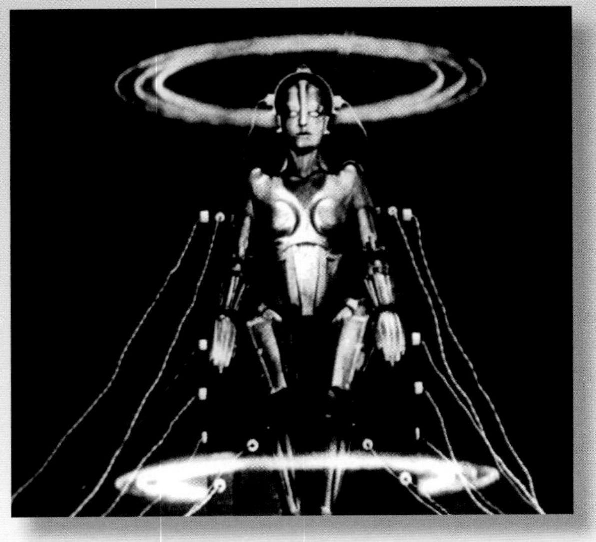

The first movie android, shown above, was featured in *Metropolis*.

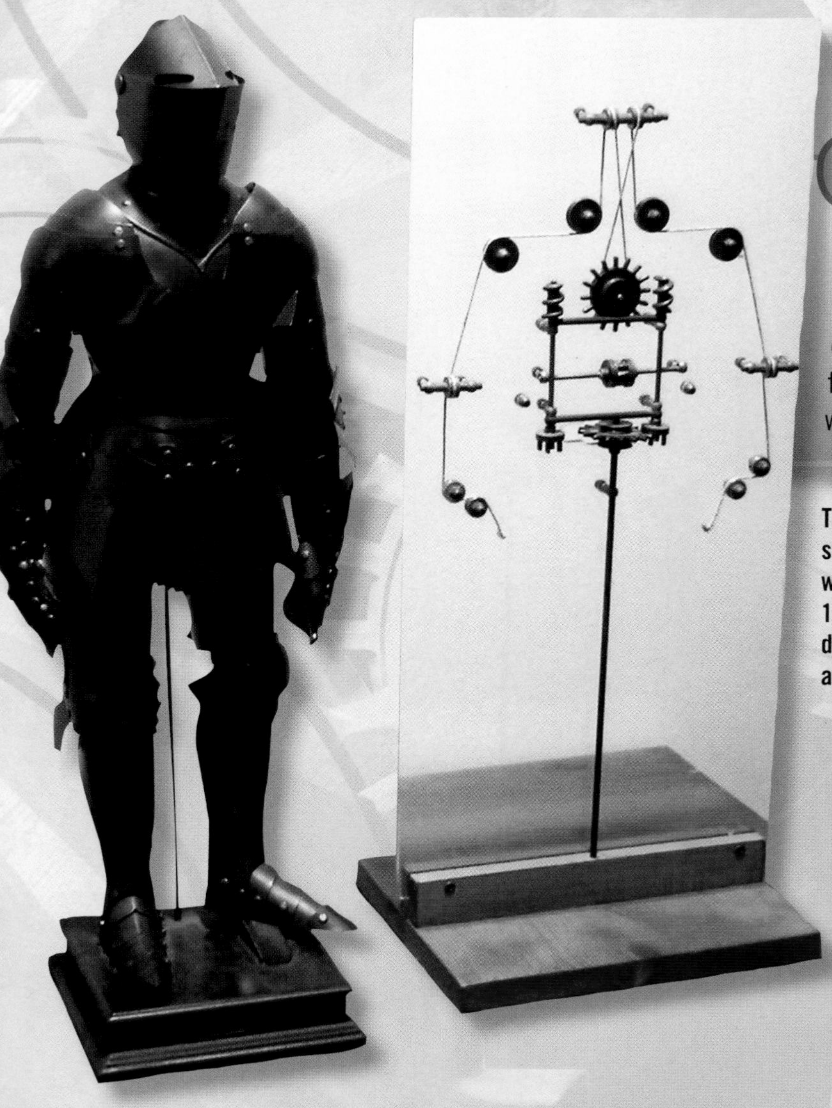

ROBOT DATA

LEONARDO'S DESIGN may be the first modern conception of a robot. His notes and drawings for the automaton, another term for "robot," were not discovered until 1950, when the device was built as a curiosity.

This robotlike knight, shown with its inner workings, was built in 1950. It is based on da Vinci's sketches of a mechanical man.

1950 book *I, Robot*, Asimov featured robots that took care of children and were friends and coworkers of humans.

Whether they were harmful or helpful, the robots of Capek, Asimov, and many others remained in the world of fiction. But there were real robots on the horizon, and they would soon be ready to take the world stage.

Robot Rules

IN "RUNAWAY," a 1942 short story, science-fiction author Isaac Asimov devised three laws that he said should control the behavior of robots:

1. **A robot may not injure a human being** or, through inaction, allow a human being to come to harm.
2. **A robot must obey orders given to it by human beings,** except where such orders would conflict with the First Law.
3. **A robot must protect its own existence** as long as such protection does not conflict with the First or Second Law.

Building Robots

The 1939 World's Fair, held in Queens, New York, featured Elektro, the first humanoid robot built in the United States. Now residing in the Mansfield Memorial Museum in Ohio, 7-foot Elektro was one of eight robots built by electrical manufacturer Westinghouse between 1931 and 1940. Elektro could walk, talk, use its arms, move its head and mouth, and . . . smoke cigarettes! As for more practical purposes, however, Elektro was no match for a robot introduced some 20 years later.

Robot Pioneers

Ever since he was a young boy, George Devol Jr. was interested in electronics and machines. At the age of 20 he founded a company that produced an automatic door known as the Phantom Doorman. He went on to work for a series of companies in various capacities, but it was in 1954 when he came up with the idea for Universal Automation, which would forever change the face of robotics and factory production.

DID YOU KNOW?

A lot of industrial robots aren't very exciting—they're mostly just robotic arms with a set of mechanical muscles that work like a human arm. Robotic arms are controlled by computers and do things like weld automobiles together and make machine parts for other industries.

Westinghouse built Elektro to advertise its products, and the robot became nearly as famous as its manufacturer.

"Radio Man" WALKS, TALKS, AND YODELS

August Huber putting finishing touches on his "radio man." At left, covering is removed to show the intricate mechanism

TOWERING seven feet high, a strange "radio man" has just been completed after ten years of arduous work by August Huber, a Swiss engineer. Beneath its jointed steel body, the gigantic mechanical man is a maze of automatic switches, relays, and other controls. Microphones within the automaton's ears pick up spoken commands and carry them to an intricate system of twenty electric motors that make the fantastic creature walk, talk, sing, or yodel at the will of its master. Power for these various activities is supplied by batteries concealed in the ponderous legs. When this modern monster talks through the loudspeaker installed in its chest, its lips move in time with its speech. An ultra-short-wave receiver installed in its torso enables the "radio man" to follow orders transmitted to it by radio from remote points.

122 POPULAR SCIENCE

NEW AND IMPROVED?

IN 1939, THE MAGAZINE *Popular Science* featured Radio Man, a 7-foot-tall robot that could "walk, talk, and yodel!" The first robot with microphone ears and a speaker in his body, he moved by remote control.

From the 1920s right through the 1970s, technology magazines such as *Popular Science* and *Modern Mechanix* featured articles on plans for robots, from robot planes that would relieve pilots from air combat duty to household robots that did the laundry to robots that boxed.

14 Science and Invention for May, 1924

Radio Police Automaton

Distant Control by Radio Makes Mechanical Cop Possible

By H. GERNSBACK

Elektro was joined at the World's Fair by Sparko, Westinghouse's robotic dog.

After applying for a patent, Devol went looking for a company to finance his invention. During his search he met Joseph Engelberger, an engineer with a company called Manning, Maxwell, and Moore. Engelberger saw great promise in Devol's mechanical brainchild but, before the idea got off the ground, the company was sold and Engelberger's division closed. In 1956, the two men found a new backer in Consolidated Diesel Electronic and started the world's first robot company, Unimation. This company produced the Unimate, a robotic arm that could help build cars on the assembly line.

The original Unimate took metal parts from the assembly line and welded them onto larger automobile bodies.

Practical Robots

Once the robot revolution took hold, there was no stopping it. Slowly but surely, industries and companies big and small found use for these mechanized helpers. Today, robots have found their way into virtually every corner of our world.

Where once men and women populated assembly lines, much of the work is now done by robots and monitored by humans. While modern medicine is still in the hands of skilled doctors and surgeons, robots now make it easier for them to perform their jobs. The same goes for military and law enforcement sectors, where robots often save lives.

Whether they are scouring the depths of the ocean floor or vacuuming your apartment floor, robots have still only just scratched the surface of all of their practical applications.

Research Robots

Robot research involves a lot of trial and error to see what works and what doesn't. That's where specialized robots come into play for use in laboratories. Research robots can help scientists learn how to design better robots or figure out how robots and humans can interact more naturally.

Loosen Up

Kotaro is a Japanese robot with a sophisticated system of artificial muscles and joints whose specialty is movement. Kotaro is trying to mimic the flexibility and suppleness of the human body so robots can dump the Frankenstein walk for which they are famous in exchange for a more natural stride.

Meanwhile, another robot named Isaac is working on how to keep one foot on top of a ball without tipping over, a balancing act that may help him climb stairs and perhaps strike a pose for some future robot soccer team.

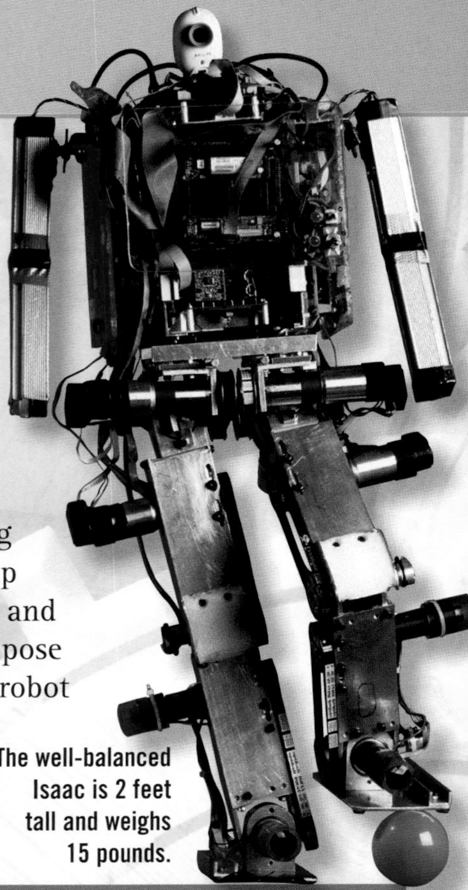

The well-balanced Isaac is 2 feet tall and weighs 15 pounds.

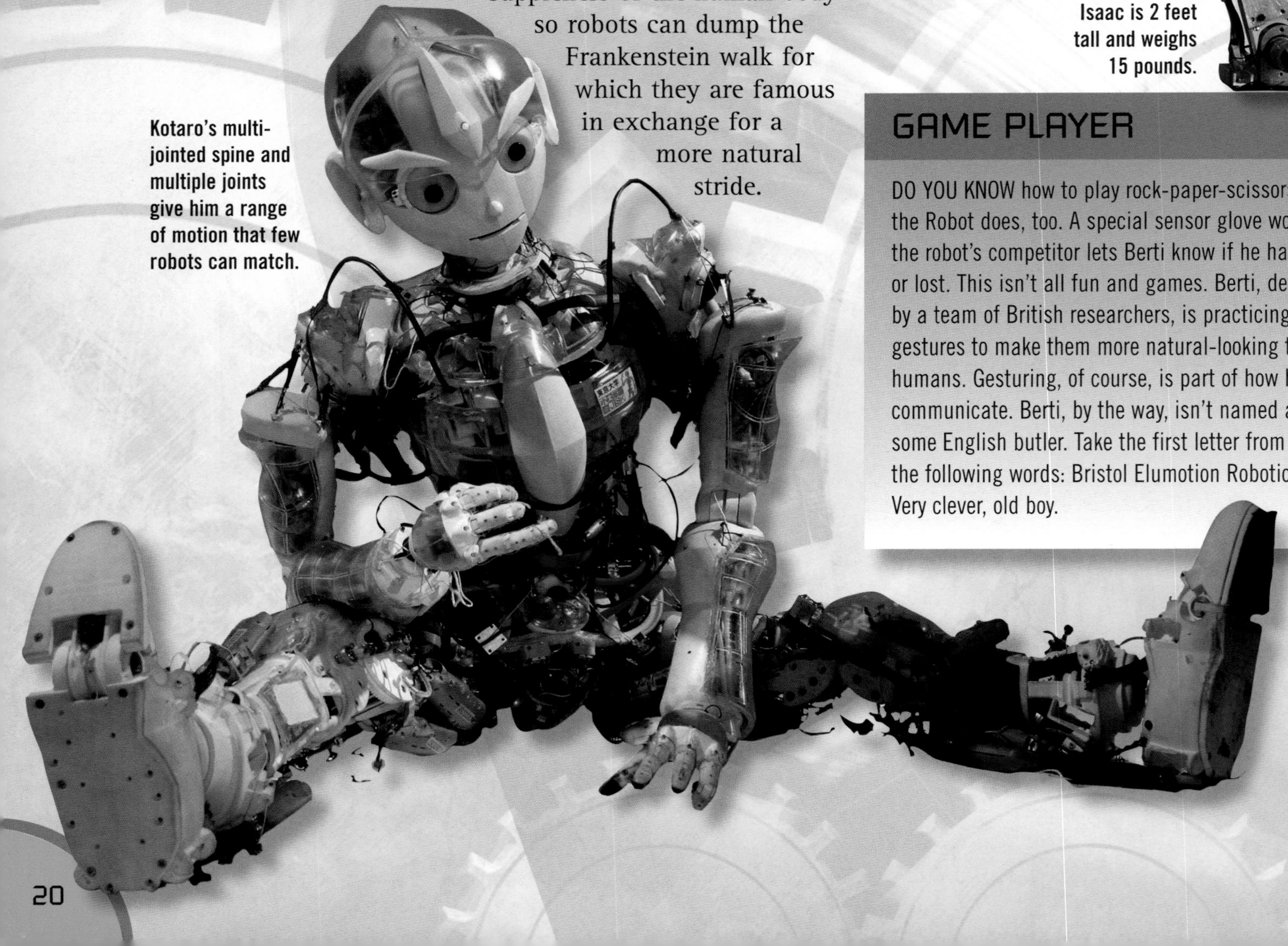

Kotaro's multi-jointed spine and multiple joints give him a range of motion that few robots can match.

GAME PLAYER

DO YOU KNOW how to play rock-paper-scissors? Berti the Robot does, too. A special sensor glove worn by the robot's competitor lets Berti know if he has won or lost. This isn't all fun and games. Berti, developed by a team of British researchers, is practicing his gestures to make them more natural-looking to humans. Gesturing, of course, is part of how humans communicate. Berti, by the way, isn't named after some English butler. Take the first letter from each of the following words: Bristol Elumotion Robotic Torso 1. Very clever, old boy.

Face Time

Other research robots are using an array of sensors to identify human emotions and then mimic facial movements associated with them so that they can raise an eyebrow in surprise or effect any number of other expressive poses.

Not all robot research is aimed at robots, though. Italian scientists at the University of Genoa use a Babybot—essentially an oversized head stuffed with circuitry, gyroscopes, and sensors—to study how light and sound affect human brain development.

IT'S KISMET!

THE MASSACHUSETTS INSTITUTE OF TECHNOLOGY (MIT) developed a robot named Kismet, which resembles a human in many significant ways. Kismet can see with its camera eyes, hear with a microphone, think with its nine-computer-strong brain running Linux software, and move realistically with its 21 articulating motors.

Kismet responds to human facial expressions and movements by moving its eyebrows, eyelids, lips, and jaws to let you know if it's happy, angry, sad, surprised, or even disgusted.

Kismet can also learn from its human encounters because its software gets data from the microphones and cameras in real time to recognize and mirror the tone of human speech or the crinkling of a brow. It can duplicate human facial expressions and, like military robots and chess-playing robots, anticipate actions.

Once you get past its exposed hardware, Kismet displays a wide range of human emotions.

Industrial Robots

Robots, with their ability to perform the same movements over and over without getting tired or bored, are ideal for certain kinds of jobs, such as working on assembly lines in factories. And following in the footsteps of Unimate, the first industrial robot, that is just what most robots that exist today do: they work in industry, building many products that most of us can't live without, such as cars and trucks.

Industrial robots excel at tasks that can be very hazardous for a human, such as welding metal parts together.

Factory Skills

There are many kinds of tasks that industrial robots must perform with speed and precision, including

- welding
- painting
- ironing
- assembly
- pick and place
- packaging
- product inspection
- product testing

A robot welder is used to assemble the gigantic pipes that move oil from one place to another.

Types of Industrial robots

There are several kinds of robots that work in factories, but two that are very commonly used are articulated robots and SCARA robots.

Articulated Robots. An articulated robot has joints that rotate. The joints are arranged in a "chain," so that one joint supports another farther along in the chain. The rotating joints allow them to pick up things and move them around, screw and unscrew bottle caps, swirl liquids, paint and spray, drill, inspect, flip switches, dispense liquids and other items, punch buttons, and perform many other useful tasks.

SCARA robots. SCARA stands for Selective Compliant Assembly Robot Arm or Selective Compliant Articulated Robot Arm. They are a lot like articulated robots, but the jointed two-link arm layout is similar to our human arms. They are often used to assemble things.

Other Uses

Most industrial robots make things, but they can do other things, too. For example, automated guided vehicles, or AGVs, carry material from one place to another in a factory. Controlled by a computer, AGVs follow a wire embedded in the factory floor to get from place to place.

Top: An articulated robot arm. This one is used to screw parts into place. The shape and structure of the tip of the arm, or "hand," varies depending on the robot's task.

Above: A SCARA-type robot working on an assembly line

Left: An articulated robot picking up bricks on a conveyor in an automated brick-making plant

Robots in Medicine

Robots, with their ability to maneuver easily in places humans can't, make useful medical assistants. Surgeons can now operate without even touching a patient, using a multiarmed robot surgeon.

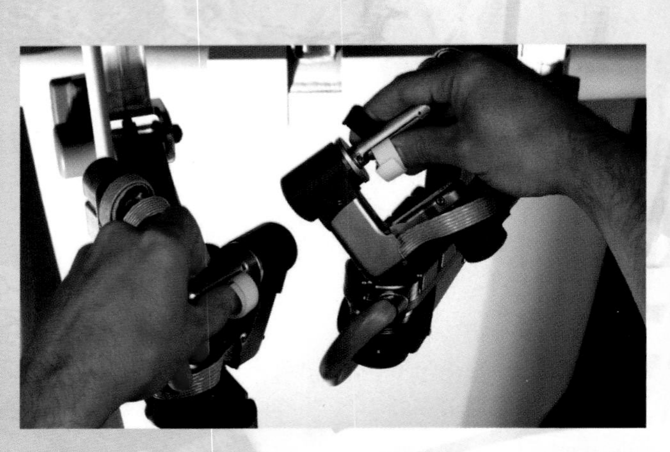

The da Vinci robot system allows a surgeon to operate its robotic arms from a computer console located a few feet away from the actual operating table.

Robot Surgeon

At the Vattikuti Urology Institute at the Henry Ford Hospital in Detroit, Michigan, doctors use the controls of a computer to move the tips of micro-instruments. The techniques allow the surgeon to operate through several dime-sized incisions—and smaller holes means faster healing and less time spent in the hospital.

A Modern Da Vinci

With the da Vinci Robotic Surgeon System, developed by Intuitive Surgical, Inc., surgeons should soon be able to operate on the heart without stopping its beating. And medical robots aren't just for surgery. British inventor Alex Zivanovic has developed a medical robot that gives shots, ensuring that every injection finds its target.

In some hospitals, PatrolBot and Robo Courier navigate hallways on their way to pick up or deliver patient samples, such as blood and urine. The self-navigating utility robots remove the grunt work of distributing and collecting patient samples, freeing trained medical personnel to do more important work.

The da Vinci system is capable of memorizing the position of its robotic arm so that instruments can be changed with little disruption to the surgery.

PRACTICE MAKES PERFECT

PRACTICE MAKES PERFECT at the National Autonomous University of Mexico in Mexico City (UNAM), home to the world's largest robotic training center. Students can deliver a robotic baby (from a robotic woman), revive a robot who had a heart attack, or perform a standard examination of robot patients. This doesn't replace examining real patients, but the constant repetition will help students improve their skills. And if somebody messes up, it's pretty easy to push the reset button!

OUCH!

IN JAPAN, Nippon Dental School in Tokyo has helped to develop a dental training robot android that can "feel pain." The robot alerts the dentistry students if she's uncomfortable. With her skin on, the Simroid robot (Simroid is short for "simulator humanoid") looks like a real woman, but her full set of white teeth is fitted with sensors. Pain Girl, as she's known, yelps when the dentist's equipment touches the virtual nerves. When a dental student touches the mouth of the Simroid incorrectly, she responds with: "Ow, that hurt!"

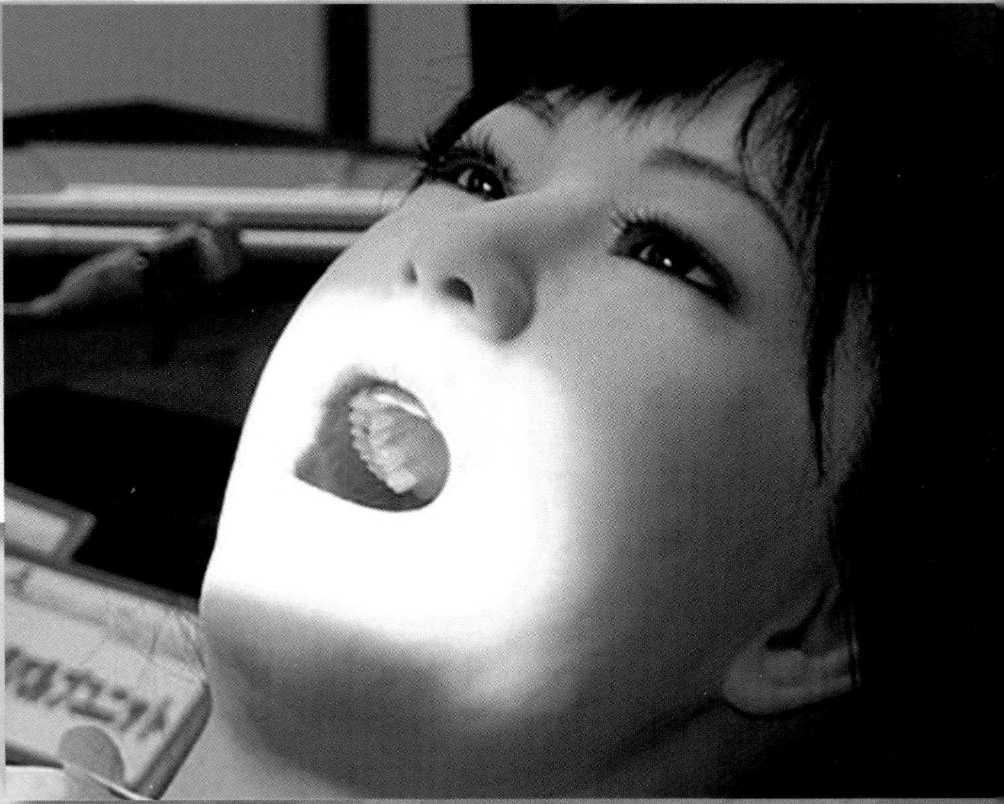

Site of Nippon Dental School, where dental students may practice with medical equipment on robot patients

Site of UNAM University, world's largest "robotic training center," where medical students can practice

Detroit, Michigan

Mexico City, **MEXICO**

Tokyo, **JAPAN**

Site of Vattikuti Urology Institute, Henry Ford Hospital, innovator of robotic surgery

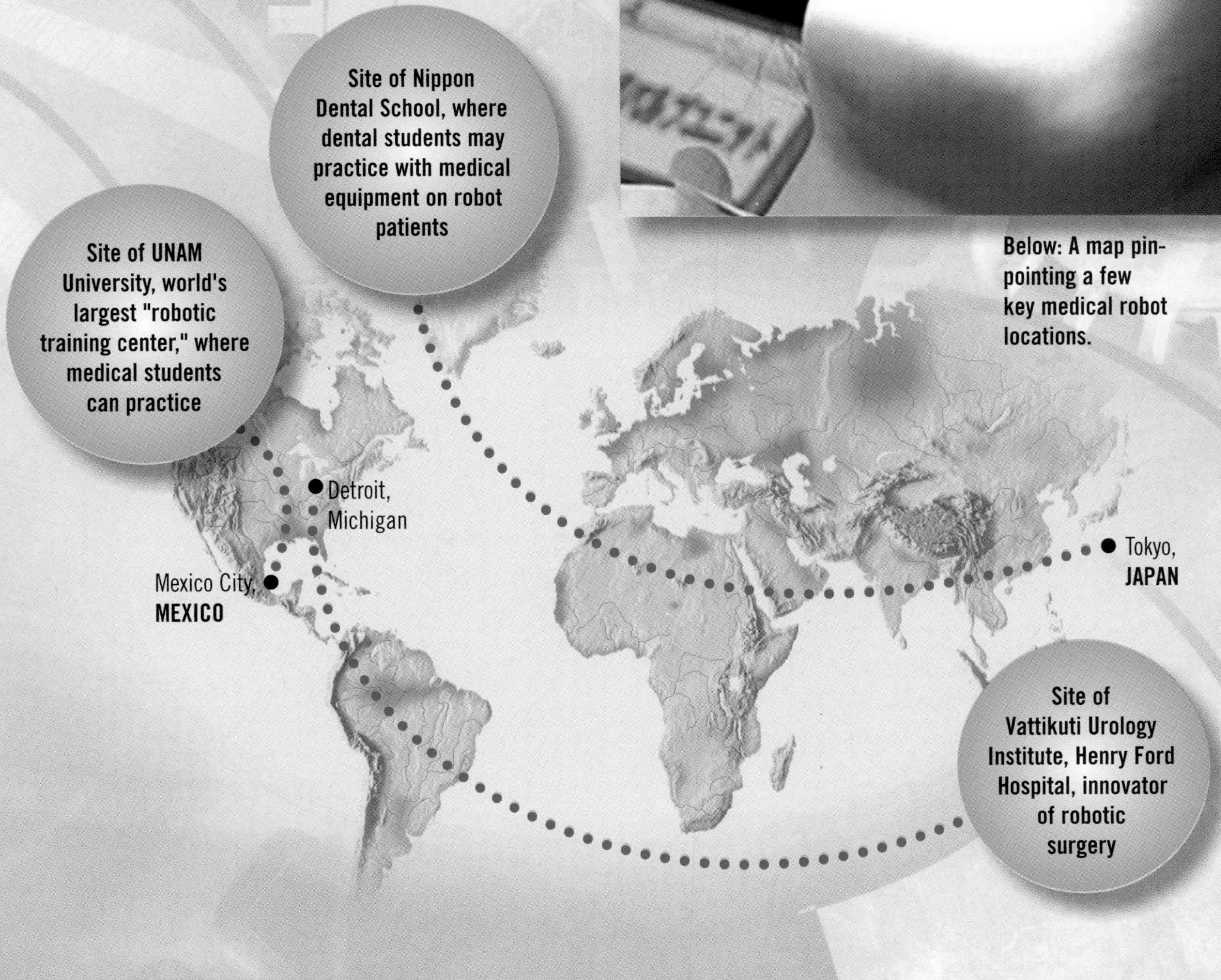

Below: A map pin-pointing a few key medical robot locations.

Above: Not just a pretty face, the Simroid robot Pain Girl communicates with dentists by using arm and hand movements, as well as speech.

Military Robots

The science of military robotics is a relatively new one. Though the U.S. military, for instance, was developing robotic systems in the 1950s, most of the robots that came out of these programs were used only for taking pictures of enemy targets.

Times have changed. These days, some robots even work on the front lines in Iraq. New military robots can walk through minefields, deactivate unexploded bombs, or clear out the enemy from hostile buildings.

These humanoid robot soldiers are still many years away from becoming a reality.

SWORDS and the Warrior

An early effort in robotic soldiering was a small tracked vehicle called SWORDS (Special Weapons Observation Reconnaissance Detection Systems) that could employ a machine gun. The robot's primary mission, however, was to gather information about an enemy using a variety of sensors.

SWORDS, or Special Weapons Observation Reconnaissance Detection Systems, are military robots built with machine guns.

ROBOT SOLDIERS

ROBOT SOLDIERS ARE VERY USEFUL on dangerous missions where the risk to human life is especially high. But the idea of a robot soldier being able to fire a weapon on its own, Terminator-style, is a controversial one. Whether computer software will ever be able to independently and reliably tell the difference between friend and foe remains in the far future. Even then, a robot soldier may not look humanoid.

The iRobot Company is designing the Warrior 700 robot for the U.S. Army. The Warrior 700 model can fight fires, diffuse bombs, and carry up to 150 pounds. Since 2001, an army of 5,000 bomb-detecting robots has already discovered and defused more than 10,000 roadside and hidden bombs in Iraq and Afghanistan.

In the future, we may send robots to do even more jobs. One day, entire armies might be made of robots! In 2008, the Pentagon budgeted $2 billion for military robot research and development.

Brain gets a kick out of anticipating soccer moves.

DID YOU KNOW?

As he plays soccer, a robot named Brain (built at Carnegie Mellon University) figures out how his opponent is going to move by remembering his previous maneuvers. The military uses the device to test how robots can learn the enemy's strategy in combat.

MULE DOGS

OUT IN THE FIELD, the average soldier carries from 50 to 90 pounds of gear. The BigDog robot, currently in development, will help soldiers carry even heavier loads. BigDog can heft up to 340 pounds of supplies while walking, running, jumping, and climbing slippery ground, snow, or rubble—places that a jeep might not be able to go. Its "knees" bend inward, and it looks kind of buglike (though it has only four legs). Sensors help it navigate and keep its balance.

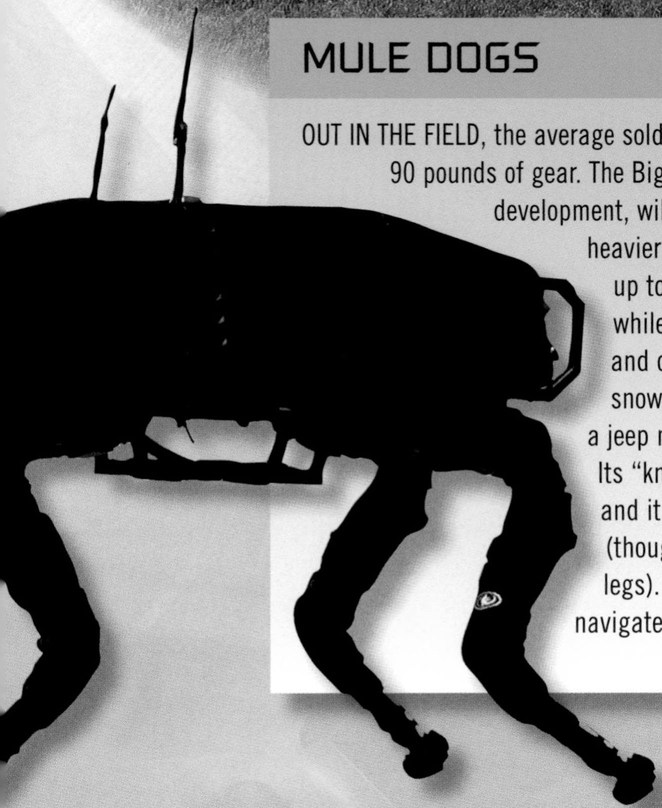

BigDog is made by Boston Dynamics, a robotics company that specializes in military robotics.

Robot Planes

Take the pilot out of the plane and you have a military aircraft that can fly dangerous missions at no risk to human life. That's the idea behind unmanned fighters being developed by airplane manufacturers like Boeing. There is a performance benefit as well—unmanned planes can make maneuvers that would cause a pilot to black out. This means an unmanned aircraft may be more able to survive enemy attacks.

Tiny Aircraft

Often referred to as unmanned aerial vehicles (UAVs), these flying drones come in all sizes and include helicopter models. There is even a new class of UAVs called Nano Air Vehicles because of their tiny size. One model, called the Katana, is a single-bladed helicopter about the size of a coin. It will be used for indoor spying missions and can be launched just as you might toss a paper airplane. As useful as UAVs are, there are still some challenges. Primary among them is getting a squadron of drones to fly together in formation without crashing into one another. Another is being able to control a drone from another flying aircraft nearby.

UAVs are increasingly being thought of as tools for nonmilitary missions like monitoring forest fires. And the day may soon come when large drone aircraft will ferry cargo from coast to coast.

PREDATOR

THE MOST SUCCESSFUL drone aircraft is called the Predator and is now routinely used in military operations. It can be equipped with a variety of cameras and sensors, including synthetic aperture radar (SAR) that allows it to easily see through clouds and sandstorms. Predators are sometimes armed with Hellfire missiles or laser-guided bombs for use against enemy targets. The Predator can fly to an altitude of 50,000 feet and has been in use since 1994. The aircraft is 27 feet long and has a 49-foot wingspan.

Some unmanned air vehicles, like this Microbat, are so small that they can almost fit in the palm of your hand.

Above: Like most UAVs, the Predator drone is operated by its pilot from the ground via remote control.

Below: Global Hawks are used primarily for military surveillance.

DID YOU KNOW?

The long-distance flying record for an unmanned plane was set by the Global Hawk in 2001. It flew from California across the Pacific Ocean to Australia, a distance of 7,500 miles. The flight took nearly 24 hours to complete. Global Hawk can operate independently and can file its own flight plans. Originally designed as a spy plane, Global Hawk is now also used to monitor climate change high in the atmosphere. Global Hawk can fly to an altitude of 65,000 feet and stay aloft for 31 hours. More recently, a hand-launched solar-powered UAV called Zephyr claimed to stay airborne for 83 hours and 37 minutes.

Law Enforcers

How do you train law enforcement officers to identify and shoot bad guys while avoiding shooting innocent passersby? To train their officers, some police and other law enforcement agencies use robots for target practice.

The "Mad Robot"

United Service Associates makes the "Mad Robot," which is basically the crash-test dummy of law enforcement. This robot, a mannequin target on remote-controlled wheels, makes sudden and unpredictable movements in training scenarios. This way, police officers can learn not to react with surprise. Instead of being startled, they learn how to blast the robot.

Bomb Sniffer

Mini-Andros, made by the Remotec Corporation, has all the features a good bomb disposal robot should have. Its cameras can see in low light, its tracks can climb stairs, and its robotic arms can pick up a bomb and move it to a remote location, where it can be safely destroyed. Robots like these save the lives of human bomb defusers every week.

The Mad Robot is programmed to strike out any which way. He runs on battery power and can be controlled by a remote human operator.

Jelly Robot

The U.S. Defense Department is currently developing a "chemical robot" made of soft, flexible materials that can compress, allowing the robot to squeeze through small openings. Then the robot reconfigures itself to do its work. Robots can disarm bombs or see around corners or up stairs, but their size and shape can limit where they go. A flexible robot would be able to get into those hard-to-fit places so that an actual human doesn't have to do the dirty work.

TWO-PLANET ROBOT

THE SANDBOT ROBOT designed for Mars exploration shares technology with RespondBot. This helpful Earth-based robot helps police and other officials face situations too dangerous for humans to handle. The robot can enter burning buildings, show rescuers if the premises are safe enough to enter, and seek out trapped victims.

The RespondBot has six curved legs that rotate like wheels and allow it to handle almost any kind of terrain. It can even climb stairs.

This robot, called a robot zapper, is used by police to take care of dangerous explosives.

DID YOU KNOW?

Stakeouts—when police secretly watch a criminal's movements and activities—can be pretty boring for officers, who may wait around for hours or even days for something to happen. Small spy robots, used for surveillance, have video cameras to record all the action. This robot can also find suspects hiding in vents or drainage pipes. Like the bomb disposal robot, the surveillance robot is controlled by a remote operator. Robots can check areas for criminals before humans rush in. In a stakeout, robots could watch a building or situation for suspicious activities or even sense heat patterns to tell where anything warm-blooded might be hiding.

The robot miner rolls out, ready for deployment as an expert bomb sniffer and defuser.

Robots in Space

Humans dominated the early history of space exploration, but it's safe to say that in the future there will be more robots in space than astronauts. Robots are already exploring the surface of Mars, and they will be key players in future space missions to the moon. And given the high radiation levels, the lack of water and oxygen, and other hazards, deep space exploration will likely be left to robots for the foreseeable future.

Robotic planet explorers may look and move differently than we anticipate. For example, the Jollbot, developed at the University of Bath in England, is a 2-pound sphere that rolls across smooth terrain. When it encounters an obstacle, however, it mimics the jumping ability of a grasshopper to leap over it.

Robonaut is a humanoid robot that NASA hopes will one day be able to aid astronauts in space.

Robot Independence

With more robots in space than people, it's likely that robots will begin to operate more independently. The main reason for this is the long time it takes to send a command across the vast distances of space. It would take over an hour for a signal to reach a robot on a mission to the moons of Saturn or Jupiter. In the movie *The Matrix*, the evil Agent Smith says: "Never send a human to do a machine's job." That might turn out to be good advice.

Astronaut Stephen Robinson hangs out on the International Space Station's robotic arm.

Opposite: A pair of astronauts share a space walk outside the Space Shuttle *Discovery* as one stays anchored to its robotic arm.

SPACE STATION'S ROBOTIC ARM

PERHAPS THE MOST FAMOUS robotic arms in the solar system are the ones aboard the United States Space Shuttle and those attached to the International Space Station. The arm aboard the Space Shuttle was instrumental in constructing the space station, and its usefulness did not go unappreciated. The Space Station is adding robotic arms as fast as they can be built. In addition to one existing arm, a second arm is being added to help maintain the Russian segments of the station. A third arm built by Japanese scientists will help conduct exterior experiments and maintain an array of telescopes. Robotic arms also serve as scaffolds for astronauts who need to be moved from spot to spot outside the hull, reducing the amount of time spent on space walks.

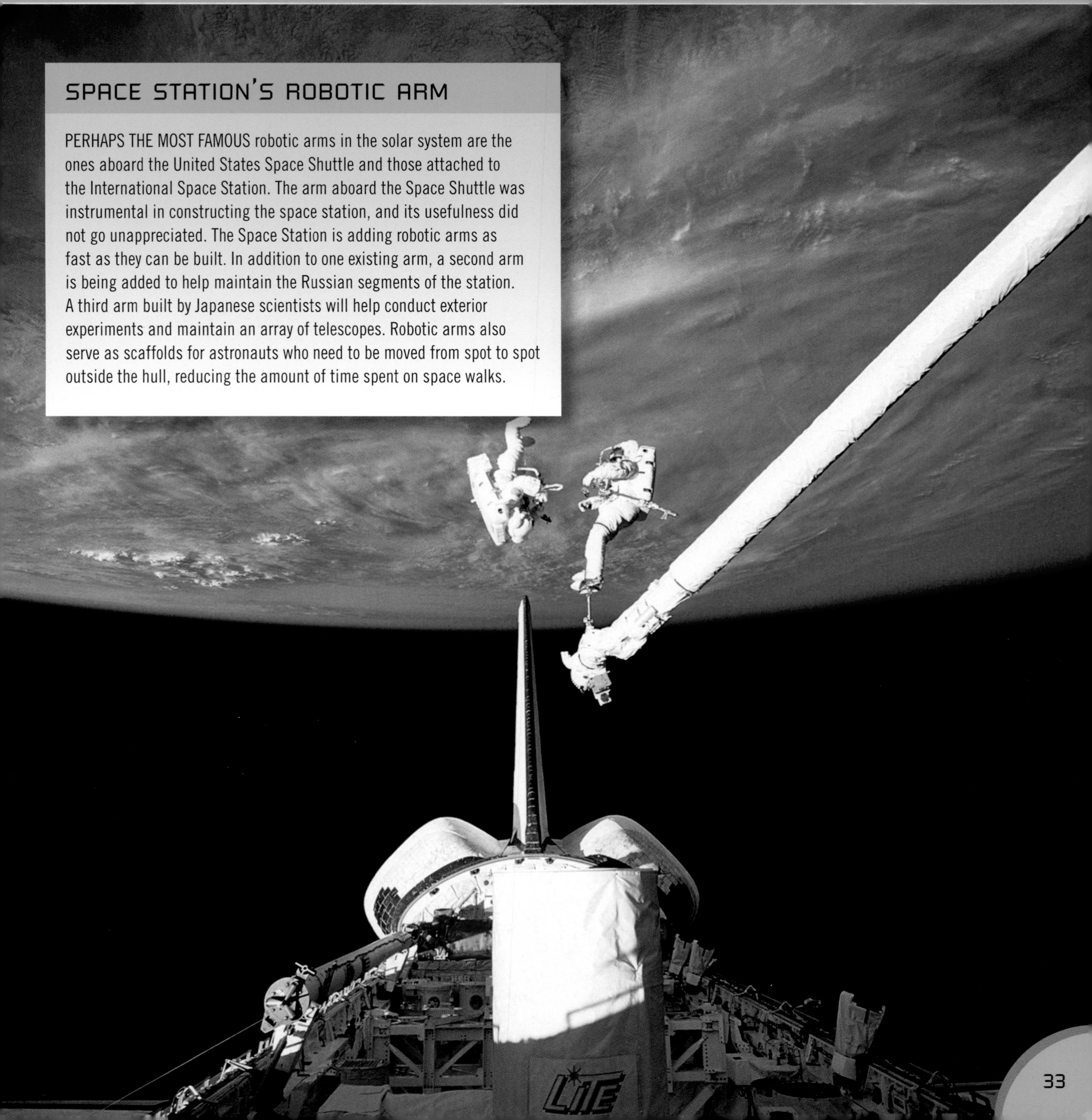

Robots in Space continued

Robots on Mars

One of the most exciting space program stories in recent years came when NASA scientists safely landed two rover robots named Spirit and Opportunity on the Martian surface. The twin robots landed on Mars in January 2004, on missions originally planned to last 90 days. The rovers astounded NASA staff when they kept functioning years beyond their expected lifetimes. They will probably keep working through 2009, though Spirit has recently had trouble absorbing the solar power it needs to keep functioning, because of fierce Martian dust storms.

A rock formation on Mars photographed by the Spirit rover. The rock in the center was named "white boat" for its color and odd shape.

An artist's painting of what the Spirit and Opportunity rovers look like as they roam the Red Planet.

Below: In November 2007, Spirit captured a stunning panorama of Mars.

SPACE BALLS

A POSSIBILITY for future Mars exploration doesn't look like your typical vehicle or robot. Groundbot inflatable rovers—big, black spheres powered by solar panels—could go more than 62 miles on a single charge. Since the balls inflate after landing, they would take up half the room in the capsule of a standard ground rover. The European Space Agency (ESA) originally planned the inflatable rovers for its BepiColombo Mission to Mercury in 2013. Though there's no room on ESA's ExoMars rover in 2013 for Groundbots, Rotundus (the rover's manufacturer) hopes that ESA will find space for the inflatables on another mission.

Signs of Life?

The rovers each carry a set of sophisticated tools to examine the geologic history of Mars. The robots have also found evidence that the area of Mars that they are exploring stayed wet for an extended period of time long ago. This water may have allowed life to develop on Mars, though that life may only have been microorganisms. Because Opportunity stalled in the sand dunes of Mars, the University of Pennsylvania developed SandBot, a six-legged robot investigator that will hopefully be able to walk in deep sand and operate in Martian sandstorms.

SandBot's ability to make its way through deep sand may come in handy if it goes to Mars.

This illustration shows what the Phoenix lander would look like, fully operational, on the surface of Mars.

Tastes Like Perchlorate

NASA'S FIRST lander robot, Phoenix, was designed to study the history of water in the Martian arctic's ice-rich soil and to explore whether the planet could support life. The robot identified water ice in soil samples and has detected the chemical perchlorate in the soil, a sign that liquid water has been on Mars in the past. Since then, Spirit has found evidence that water in some form has altered the mineral composition of soils and rocks. Before humans get on a spaceship and go exploring Mars, more robots will definitely be making additional landings to one of Earth's closest neighboring planets and the one most likely to support life.

Back To The Moon

The moon's rough terrain and lack of gravity make for a difficult environment for astronauts and robots alike.

ROBOTS INVADE THE MOON! That might be a headline splashed across newspapers sometime during the next decade as the United States and other countries develop new moon missions. The moon may serve as a launching pad for the exploration of other planets. It might also be the source of important new resources like helium 3, a nonradioactive power source for nuclear fusion reactors that is rare on Earth but thought to be abundant on the moon. Scientists estimate that 100 tons of helium 3 could power Earth for a year.

Robots will play a key role on the moon since they can operate well in environments that lack an atmosphere. One important assignment is to look for water at the moon's poles. Water is critical for long-time survival on the moon and may also be used to make rocket fuel for return trips to Earth or voyages to Mars. Robots also will be looking for other valuable minerals.

HEAVY LIFTER

A SIX-LEGGED ROBOT named ATHLETE (All-Terrain Hex-Legged Extra-Terrestrial Explorer) will get to do all the heavy lifting on the moon. ATHLETE can roll across lunar plains or walk up the sides of steep craters. Developed by NASA's Jet Propulsion Laboratory, ATHLETE can carry large payloads and can work with other ATHLETES to lift large objects like habitat modules into place. ATHLETE moves at a brisk 6 miles per hour and responds to both voice and gesture commands of astronauts. It's also equipped with a launchable grappling hook for rappeling down steep slopes. ATHLETES can be stacked on top of each other for easy transportation and storage.

Future models of ATHLETE may move up to 100 times faster than the rovers currently exploring Mars.

Rovers like the Scarab are put through all kinds of rigorous tests, including simulated sand storms and extreme temperature changes, before they are sent into space.

Prospector

The Scarab is a robot prospector that would roam about the moon's south pole looking for valuable minerals to mine. Human prospectors used to look for gold with pickaxes and their bare hands, but Scarab will be better equipped. Laser scanners will help it navigate harsh terrain and a radioactive isotope should keep it running for 10 years, say developers at Carnegie Mellon University. The Scarab flops down on its belly to drill for samples.

PREPARE FOR LANDING!

A PAIR OF ROBOT CONSTRUCTION WORKERS the size of riding lawnmowers may pave the way for later astronaut landings by building a hard surface landing strip with a surrounding wall. This would protect future crew quarters from the sandblasting effect of rocket blasts during takeoffs. Developed by Astrobotic Technology Inc. and Carnegie Mellon University researchers, the two 330-pound rovers are essentially a shovel and dump bed mounted on tracked wheels. The two moon diggers would complete the job—which involves moving 2.6 million pounds of lunar dirt—in about six months.

Robots Underwater

The earth's oceans are difficult to explore. Deep down, the water pressure is crushing and it's totally dark: no person can hope to dive to the ocean's floor. In fact, although almost three quarters of the planet's surface is covered with water, we have yet to investigate 95 percent of what's beneath the surface. Robots may soon change all of that.

Self-sufficient Sub

One underwater vehicle that's already made a big splash is the Autosub. Launched in 1996, Autosub is an independently operating mini-submarine capable of extended trips under the ice of the Antarctic, where it has a fish's-eye view of oceanic activity. Once its data is gathered and mission is complete, Autosub returns to its ship.

DID YOU KNOW?

Instead of churning up the water with thrusters or motors, the underwater explorer robot Aqua uses its six fins to swim like an animal. Aqua can also walk on the ocean floor, swim on the surface, or climb and inspect coral reefs, which can provide clues to the ocean's health. Its camera eyes help it navigate as its computer brain figures out the best route. Based on a six-legged robot originally developed to explore Mars, amphibious Aqua can also walk on land.

Above: The brutally cold waters beneath the arctic ice shelves prove no problem for the Autosub.

Left: Aqua robot swims along the seafloor. It is very similar in design to both SandBot and RespondBot.

RoboShark

IN 1989, OCEANOGRAPHER Jacques Cousteau and his son Jean-Michel created a robotic shark with a built-in camera. They wanted to get an up close and personal look at how real sharks act when they don't know that anybody is looking. Cousteau's grandson Fabien continues the family tradition of deep-sea exploration with a submarine that moves like a great white shark to better study these mysterious creatures.

Fabien Cousteau describes his robot shark submarine as "a 1,200-pound tool that looks, feels, and moves like a great white shark."

Coral Explorers

In England, Durham University scientists are using robots similar to the Mars rovers to explore undersea volcanoes, more than 2 miles below the surface. Lophelia II 2008, an undersea robot mission, seeks out deepwater coral reefs and shipwrecks. This mission, carried out by the National Oceanic and Atmospheric Administration (NOAA) is part of a four-year project to explore new deepwater coral communities at both natural and human-made reef sites. Similarly, the robot SeaEye Falcon will search for new coral sites in the deep Gulf of Mexico.

Remotely operated vehicles (ROVs), such as this one, can photograph shipwrecks and other underwater curiosities while transmitting those images directly to their ship.

Robot Helpers

Robots can be great helpers to humans, performing such everyday tasks as vacuuming and mowing the lawn and also more ambitious ones such as helping the blind and visually impaired lead more independent lives.

RoboMow, the robotic lawn mower

Home Helpers

There's not a robot in every home yet, but with countless technological advances being made, it might not be far off. After all, who wouldn't want a robot that can do your household chores for you?

Roomba, the robotic vacuum cleaner
Above: the Roomba at work vacuuming up dust from the floor

Hate vacuuming? The iRobot Company that manufactures military robots also makes the Roomba home vacuum robot. This is a little round robot that uses ultrasound to find its way around the house. You take it to the area that you want to vacuum, turn it on, and let it go about its work. It can sense stairs, so it doesn't fall down them, and if it gets tangled in something, it will stop and let out a mournful cry so that its owner can free it. Once its job is done, it can return itself to its home base.

Like Roomba, the Trilobite vacuum uses ultrasound technology to see obstacles and avoid them, but it can also store a digital map of the room and remember it for future assignments. This computer processing power gives

SR2 at work at the Los Angeles County Museum of Art. It will navigate through the rooms and halls of the museum, making sure everything is as it should be.

Trilobite the edge over the Roomba, which cannot store data into memory.

The outdoor version of Roomba, the robotic lawn mower RoboMow, cuts the grass all by itself. And, unlike you, it doesn't need reminding. When programmed, the robot will do its chores at the preset scheduled days and times. At the end of each mowing session, RoboMow returns to its base station for charging until its next scheduled assignment.

Guardian Robots

Lots of houses and businesses use security alarm systems to keep out burglars or to detect smoke and harmful gasses. Now robots can help with these tasks.

Some of these robots, such as Security Robot, or SR2, are already at work. SR2 guards the art collection in the Los Angeles County Museum of Art. SR2 reads the detailed maps entered into its programming to cruise the museum—without crashing into the valuables. It has detectors that sense movement, smoke, fire, gasses, and humidity.

DID YOU KNOW?

Mitsubishi has developed the Wakamaru robot, which can follow people and distinguish between different faces. It uses ultrasonic sensors to avoid obstacles and calls out to the person if it loses track of him or her. Designed as a companion and helper for older people, Wakamaru can remind you to take your medicine or keep an eye on the house when nobody is home. A temporary agency called People Staff has hired 10 Wakamaru robots to work in the Tokai region of Japan. It may be expensive to hire a Wakamaru, but it works out to cost the same as a human assistant.

Below: A cheery yellow Wakamaru robot greets a human by shaking his hand.

Robot Helpers continued

Roving Robot

Today, the WowWee Rovio mobile Wi-Fi Webcam can send pictures and sound to your PC wirelessly and operates by remote control all day, every day from anywhere in the world via your computer. Its sensors guide the robot through obstacles, and a built-in microphone and speaker allow for two-way conversation. From anywhere in the world, you can make sure your home is safe. Rovio automatically monitors his power supply and returns to his dock when power is low.

Companion Robots

At MIT, researchers are working on a very early version of an intelligent, robotic helper. Domo is a humanoid that grasps objects and places them on shelves or counters. Domo has springs in his arms, hands, and neck that can sense resistance and respond to it. A robot like Domo could help elderly or wheelchair-bound people with simple household tasks.

A Dog That Makes YOU Heel

Do you need help watching what you eat? Cynthia Breazeal of MIT has designed a robot dog that monitors what you eat by tracking an electronic food diary and communicating electronically with your scale. It pleads in English for you to not pig out. The British weekly *New Scientist* reported that the canine robot is programmed to encourage and reward people who stick to their diets. The doggie robot jumps up and down, wags its tail, and plays cheerful music while flashing brightly colored lights.

If you break your diet, the robot dog plays sorrowful music. Sensors in the robot include a pedometer (which counts the steps you take), a bathroom scale, and a personal digital assistant in which you must truthfully record what you eat each day. The data is sent to the dog by Bluetooth or Wi-Fi technology. Studies have shown that people who record what they eat and how much they exercise, and who receive encouragement, are likelier to lose weight.

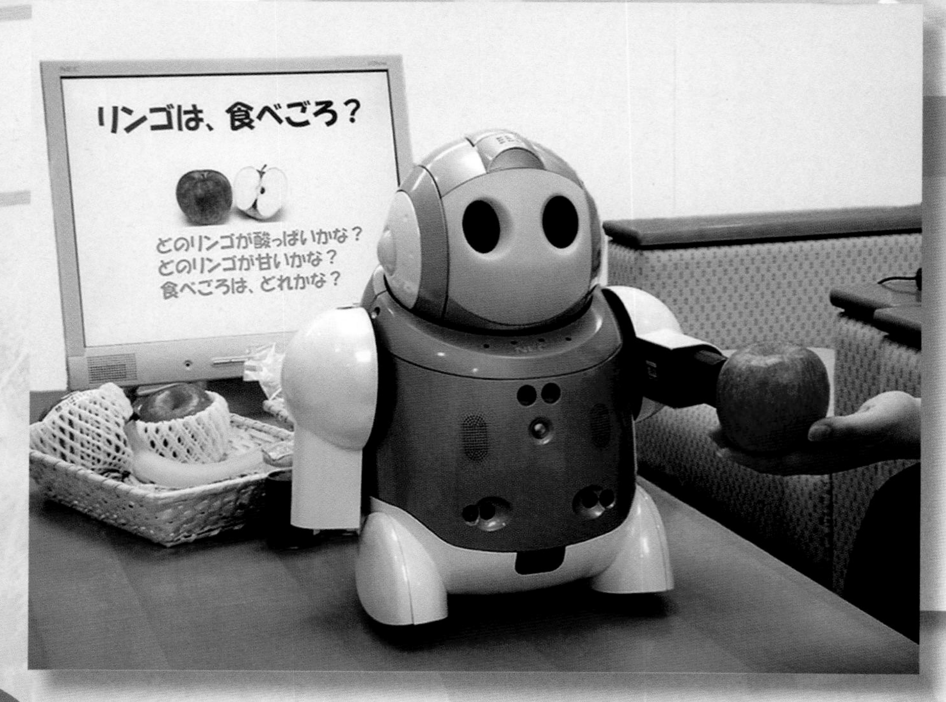

The Tasting Robot identifies an apple.

HOW TO SERVE HUMANS?

NEC SYSTEM TECHNOLOGIES invented a Health and Food Advice Robot called the Tasting Robot. It can identify types of food like cheese, meat products, and bread by shooting infrared rays at the food and then analyzing the data feedback.

According to the robot, an analysis of two different human hands indicated that people may taste like bacon or prosciutto (an Italian salted ham). NEC hopes to improve the robot so that it can help people with food allergies avoid eating things that will make them sick.

Seeing-Eye Robots

Most of us have seen someone out walking with a seeing-eye dog. These are dogs specially trained to help the blind and visually impaired. Now researchers are developing a robot that can do the same thing. Like a seeing-eye dog, the Robotic Guide, called Meldog, can help the visually impaired get around on streets and even in busy places like airports, shopping malls, and grocery stores. A guide dog will still be a visually impaired person's best friend, but a robot system will add to his or her independence.

Meldog uses a sensor that hones in on radio frequency identification tags. The tags, which can be placed in any indoor environment, tell the robot where it is. By reading a Braille directory, the user can pick a target location. Once the robot gets the user to the location, it can give information that dogs can't. For example, it can even tell the user on which grocery store shelf to find the breakfast cereal!

Meldog guides its owner through a walk in the park.

Robot Pets

Could a robot dog be man's new best friend? The notion certainly has its appeal. No need to carry a pooper-scooper. No fleas to worry about. They would also be a relief to those suffering from allergies. And studies show that robot dogs work nearly as well as real dogs in making seniors at nursing homes feel less lonely.

Sony's AIBO robot dog debuted in 1999 but was discontinued in 2006 after about 150,000 were sold—despite a high price tag of $2,000. AIBO was a major advance over mechanical dogs because it had a degree

DID YOU KNOW?

What has the speed of a tuna, the acceleration of a pike, and the navigational ability of an eel? Answer: a robot fish swimming in the London Aquarium. Developed by scientists at the University of Essex, it swims by wiggling like a real fish. Similarly, Japanese scientists have created a robot fish that looks much like a red snapper. Both fish have realistic scales. Robot fish may someday swim in your home aquarium. In the meantime, robot fish could perform tasks like exploring deep seabeds and finding leaks in underwater oil pipelines.

Sony released AIBO as a robotic pet, but it has proven to be a valuable research tool. Artificial intelligence researchers use it in experiments that work on ways to program robots to work together.

of artificial intelligence that allowed the robot dog to respond to external commands. AIBO also seemed to become more mature over time as it learned from experience.

Dino Mite!

AIBO spawned many copies of all sizes and cuteness levels. Cat lovers, of course, got equal time with various robotic felines that don't need a litter box. These purring pets, however, are no match for Pleo, the robot dinosaur. Sensitive to light and sound stimulation, this cute prehistoric pal grazes, naps, and can see with camera vision while storing the data in his computer memory.

Pleo's large head hides a camera and an elaborate network of sensors.

CUTENESS OVERLOAD

CUTER THAN YOUR AVERAGE adorable stuffed animal, the small robotic Paro looks and sounds like a fuzzy baby seal for a reason. Called a Mental Commitment Robot, Paro was developed to interact with people, giving troubled or upset people something to love. Real baby harp seals sleep most of the day, but Paro has been programmed to sleep at night and stay awake during the day, because most people sleep at night. And because the Paro is handmade, each looks a little different. About a thousand Paros have been manufactured.

Arguably the cutest robot around, Paro is modeled after a baby harp seal.

Robot Toys

Though it is often difficult to distinguish between various robot pets and toys, there is no mistaking ROBOTA. Created by Professor Aude Billard, ROBOTA dolls are a family of mini humanoid robots. They are educational toys that can engage in complex interaction with people involving speech, vision, and the ability to imitate human body movements. These dolls are very useful in helping children with autism. Their sensory systems even allow them to dance to music!

Above: Among her many talents, ROBOTA can learn and repeat short sentences in a matter of minutes.

Above, right: He may be small but don't be fooled; Cubo is one impressive little robot.

CUBO CUTIE

IN 2006, COMPANION ROBOT CUBO came on the scene. The toy-sized robot can read you audio books, teach languages, and monitor any activity in the house when you're out, recording it with a built-in camera. It has a 1.5-inch color screen and can give you news updates with its wireless Internet connection. Cubo can also stream music, play MP3s, and record voice messages. Need a wake-up call? Cubo can do it! Cubo can even retrieve weather news from the network, so you know what to wear! His camera eyes let him see where he's going and prevent him from bumping into anything.

DID YOU KNOW?

PaPeRo, a personal robot, can recognize human faces from a databank taken by its camera and can identify 10 different people at a time. With a vocabulary of 650 words, the robot can communicate simple things by speaking. The independent robot PaPeRo can roam around the room, connect to the Internet, and dance on its own. But, PaPeRo also needs attention. If you ignore it, the bored PaPeRo will become lazy and not do a thing.

Small Wonder

Zeno is another robot doll who stands tall among robots, even if he is only 18 inches high. He'll smile, laugh, recognize your face, and remember your name. Yes, he's really smart. Zeno is made by Hansen Robotics and costs about $1,500. A six-inch version may also come to market for about $300.

The Real Deal

By contrast, many other robot toys are simply dumb. They may look like robots and walk like robots but they don't think like robots, meaning there is no artificial intelligence that allows them to respond to owner's commands or learn from past experience. So don't be fooled. Real robots don't ignore humans.

PaPeRo, about 1.7 feet high and 11 pounds, is about the same size and weight as an average house cat.

TOYS, THE FINAL FRONTIER

WOWWEE ROBOTS WERE DESIGNED by former National Aeronautics and Space Administration (NASA) scientist Mark Tilden, who left the space program after a robot he designed to explore the Red Planet instead crashed on its surface. (The wreckage spread over 3 miles of Martian desert—being a robot is sometimes dangerous.)

After he moved into private business, he designed many models of WowWee robots. Controlled by remote, these can make some pretty complex maneuvers, and the highest priced ones are about $200, so a kid could show off mad math skills by pointing out that the cost is a mere tenth of AIBO—a bargain!

Roboreptile, a toy robot designed by WowWee, is a feisty, agile, dinosaur robot that you control with a remote.

Build Your Own Robot

Building your own robot is not nearly as difficult a task as you might think. You can actually create some types of robots from kits in as little as 30 minutes. Though kits vary in degrees of difficulty, some may contain a robot brain with 32-bit microprocessor power, Bluetooth, Universal Serial Bus (USB) support, and more. The robots can be programmed with either PC or Macintosh computers.

RENT A ROBOT

ONLINE STORES OFFER a wide range of educational robot kits, programmable robots, robot parts, robot books and magazines, and robot accessories. You can also rent a robot like the Sico Millennia, the most sophisticated mobile communication robot in the world, for parties and events. A very adaptable robot, Sico can change his speech, mannerisms, and behavior to get along with any age group, speak different languages, or follow any set of customs.

ROBOT DATA

SPYKEE IS A KIT ROBOT that you can program with your PC. It can be controlled remotely via the Internet using computer software. The 1-foot-tall Spykee Wi-Fi Spy Robot can walk and talk, and its webcam can perform surveillance or make videos. It also contains a digital music player for your MP3s.

Junior high school students working on their LEGO robot project.

Easy Does It

If all of that sounds a little too technical to you, don't be discouraged. There are plenty of robot construction sets around that are easier to follow. One such product is the wildly popular LEGO Mindstorms robot kit. On the market since 1998, this kit includes programmable battery-operated bricks, sensors, and gears, as well as many standard LEGO pieces. Mindstorms are currently being used in schools to help teach kids various aspects of science, technology, and even math.

Some of the parts included in the LEGO Mindstorms kit.

RoboNova comes with a rechargeable battery that offers approximately an hour of uninterrupted action.

Robots, Robots, Everywhere

Another great place to find information on building robots is on the web. In fact, there are dozens of Web sites devoted to this very pursuit. The information on these sites ranges from general advice for beginners to how to construct your very own battlebot. But, always be careful to check any information thoroughly and ask for adult supervision before going ahead with your project. When building a robot, safety is your first priority.

ROBONOVA

ONE ROBOT that has many fans is RoboNova. Teachers, students, and serious designers will love this humanoid robot. The 1-foot-tall RoboNova is capable of running, walking, flipping, and dancing. Ambitious robot enthusiasts can get this guy as a kit, and need only a screwdriver to assemble it. Younger robot fans can get it pre-assembled and ready to roam right out of the box.

Robot Games

Every year there are more than 100 robot competitions, including robot demolition derbies, robot bartender contests, and soccer matches between robot teams. There are also more "intellectual" contests, such as puzzles, chess, and other strategic games. MIT holds its 6.270 Competition every year for developers of new robot designs. The event includes workshops in designing parts and programming codes for robots.

Two robotic soccer stars square off as one takes a penalty kick during RoboCup 2006.

The RoboCup

Annually, teams of robots from many institutions take the field to play soccer against one another in an event known as the RoboCup. Competing in Cup events requires a high level of artificial intelligence and mobility. The competing robots must also be able to walk and kick a ball.

Several leagues compete in the Cup games, including divisions for humanoid robots on two legs and four-legged robots. By 2050, the Cup organizers hope to see a team of humanoid robots that can beat the best human soccer players in the world.

In 2008, about 400 teams, with 2,000 participants from 35 countries, participated in the RoboCup held in Suzhou, China!

Soccer matches at RoboCup are not limited to two-legged participants. Four-legged competitors such as AIBO also get to show off their skills.

Remote-controlled robots "Garoo" (the cute one) and "Tokotokomaru" fight it out during the 2008 Robo Japan exhibition.

Robot Combat

You might ask, why build it only to see it destroyed? But, for some robot builders, the risk of losing it all is part of the challenge. In robot combat, two or more robots, either radio controlled or autonomous, fight each other until one of them is disabled or destroyed. For a while, these contests were wildly popular, and TV shows like *BattleBots*, *Robot Wars*, and *Robotica* let viewers see the fights firsthand.

In 2002, fighting robots got their own official league, the Robot Fighting League (RFL). RFL came up with rules and regulations that organizers of robot fighting events in the United States, Canada, and Brazil must follow.

Robot Conventions

Robot groups all over the world host sporting events and conventions, where members can show off their robots' talents. There are AIBO greyhound races, in which speed gets you the prize, and sumo-style contests, in which the robot contestants compete to stay in a circle while pushing the opponent out, as well as many other games and demonstrations. At conventions like these, lots of robots get to show off their unique skills.

ROBOT RHYTHM

IT'S NOT JUST SPORTS competitions for RoboCup. The RoboCup organizers also hold a competition every year for dance. In this contest, entrants must build one or more robots that move in time to music. The robots also must dress in costumes!

A computer-controlled robotic arm makes its move against a young challenger in a game of chess.

Robots of the Imagination

As robots evolved from far-fetched dream to everyday reality, they were suddenly everywhere and capable of doing just about anything. Yet, there are still some robots that exist only in the human mind, and maybe that's not such a bad thing.

Movies are chock full of robots that can wipe out anything in their paths, such as the Terminator or the Transformers, as well as ones you wouldn't want to cross, like Robocop. And who would want to wish for a lifetime of picking up after humans, like what animated robots WALL-E and *The Jetson's* Rosie are programmed to do? It clearly isn't easy being a fictional robot.

Nevertheless, robots both real and imagined are here to stay. If you want further proof, they now have their own Hall of Fame.

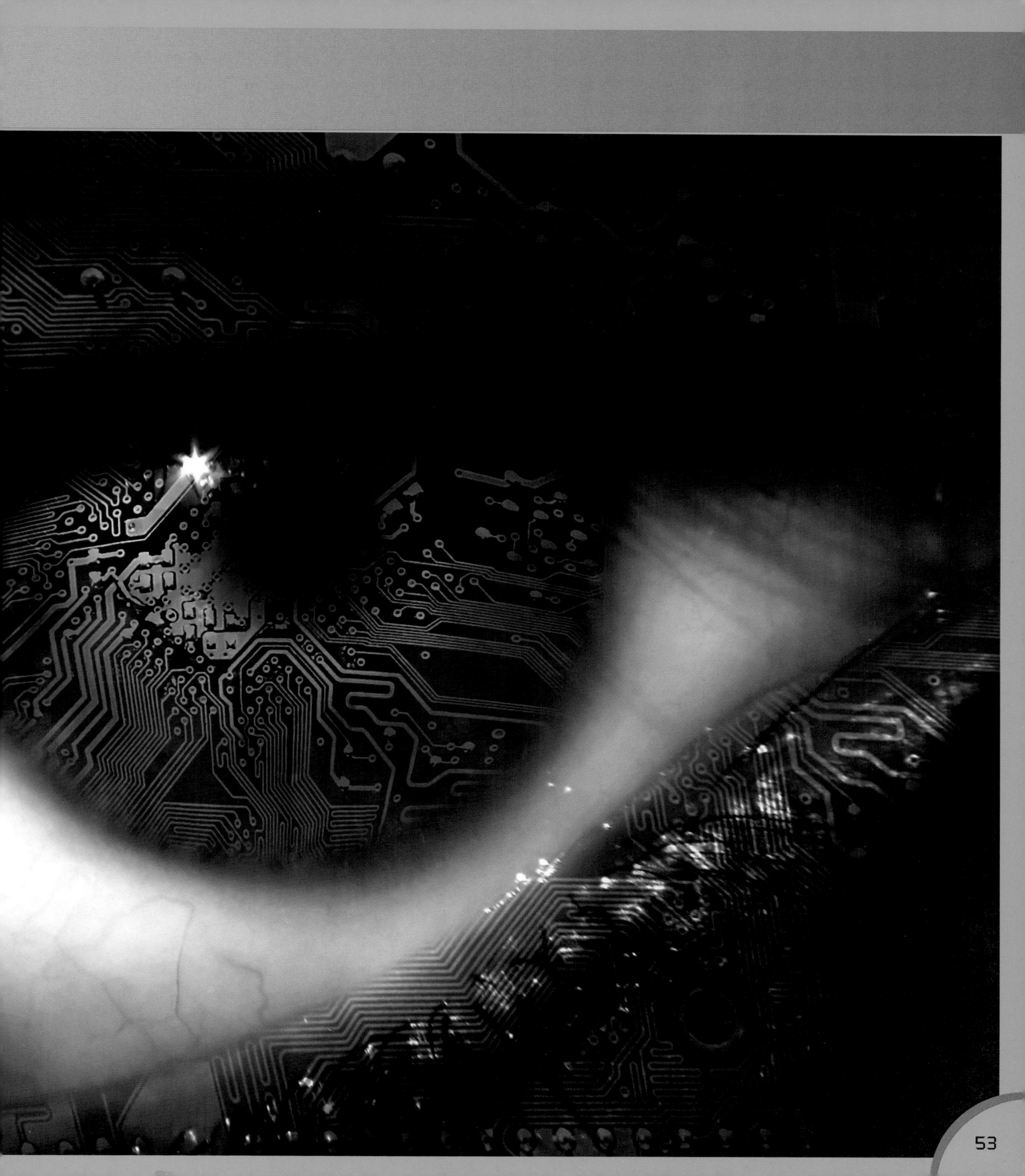

Robots on Film

Hollywood has a rich tradition of robots that have appeared on the silver screen. Like the actors themselves, these robots come in a wide variety of sizes and appearances, and play both leading and supporting roles. Their personalities range from lovable to menacing: some you want to cuddle up to and others you want to run screaming from. But, whatever parts these robots play, they almost always demand our attention.

Autobot leader Optimus Prime, from the 2007 film *Transformers*

They Came from Outer Space

Robots have always been a constant in movies about visitors from other planets. One of the best examples of this was in 2007's *Transformers*. Based on the line of Hasbro toys, the Transformers are from the planet Cybertron and capable of altering their appearances into common vehicles like cars and planes. These robots are split into warring factions; the noble Autobots and the evil Decepticons. They are all on Earth to search for an energy source to help rebuild their home planet.

A space robot with nobler purposes on our planet is Gort from *The Day the Earth Stood Still* (1957). Standing 8 feet tall, Gort's job is to police the galaxy and destroy any signs of aggressive behavior. And it's safe to say that he's not here on vacation. If the people of Earth can't stop building and using nuclear weapons Gort may just be forced to destroy the planet for the good of the universe.

Though he doesn't look particularly impressive, Gort is armed with a ray that can disintegrate weapons.

The Iron Giant's best friend on Earth is a young boy named Hogarth Hughes.

A GIANT SWEETHEART

AS SCARY AS THE THOUGHT of a colossal iron man falling from space can be, the 1999 animated movie *The Iron Giant* presented one of the most lovable aliens you would ever want to meet. Standing 50 feet tall, this robot has a childlike curiosity and innocence. He is also quite adept at repairing himself when he needs to.

Though this gentle giant has the capacity to do great damage to humans, he instead finds himself the hunted, hiding from government agents and the military. For once, a movie portrays man as being far more dangerous than machine.

Japanese superstar Godzilla goes toe-to-toe with his robot nemesis Mechagodzilla.

DID YOU KNOW?

When Godzilla stopped being a villain, filmmakers needed a new bad guy, so the Simian race from an alien planet created Mechagodzilla. Mechagodzilla had a fleshlike covering and posed as a live, breathing monster. Soon the covering was ripped off in battle, and he was unmasked as a giant robot. He has atomic breath, and missiles in his arms, toes, and fingers. He can also fire a deadly ray from his chest. He appeared in several of the Godzilla movies.

Robots on Film continued

Of course, not all cinematic robots come from outer space. In many instances they are made right here on good old planet Earth. For better or worse, these machines are the products or byproducts of human ingenuity.

Robocop's main crime-stopping competition, ED-209

BIG BAD ED

A LUMBERING MECHANICAL MENACE, ED-209 appears in all three RoboCop films as one of the law enforcement robots created by the corrupt Omni Consumer Products corporation. Heavily armed with machine guns, a rocket, and mortar launchers, this villainous robot is just about unstoppable. Designers modeled ED-209 after the Bell UH-1 Iroquois helicopter that was used during the Vietnam War.

Law and Disorder

In the 1988 futuristic crime thriller *RoboCop*, Detroit police officer Alex Murphy is gunned down and left for dead by a violent gang of criminals only to be turned into a half-man, half-robot crime-fighting supercop. With no memory of his previous life, RoboCop proves to be the best crime stopper the force has ever seen. But, when certain human traits start coming back to him, they begin to interfere with his emotionless machinelike efficiency.

On the other side of the law is perhaps the most dangerous robot in movie history, as seen in *The Terminator*. Sent from the year 2029, by a planet now ruled by computers, the Terminator is a cyborg killing machine searching 1984 Los Angeles for a woman named Sarah Connor. The Terminator's steel skeleton is covered by synthetic flesh that can sweat, breathe, and bleed like real skin. This robot assassin also has a great capacity for learning, cannot be stopped by conventional weapons, and has a power supply that can last up to 120 years. In short, it's your worst nightmare.

RoboCop is actually a cyborg, which means that he has both robotic and human elements.

Stripped down to his shiny chrome frame, it's easy to see why the Terminator is all but indestructible.

WILD, WILD *WESTWORLD*

BEFORE HE BROUGHT dinosaurs back from extinction in *Jurassic Park*, author Michael Crichton wrote and directed *Westworld*, a sci-fi thriller surrounding a robot-populated amusement park.

Within this amusement complex are three distinct destinations—RomanWorld, MedievalWorld, and WesternWorld—where high-paying thrill seekers can live out their fantasies. The androids that populate these parks are programmed to give the guests whatever they want, including gunfights in the popular Western-themed venue. But when a virus infects the massive computer that programs the robots, they begin acting irrationally and start killing off the bewildered guests. With park technicians helplessly trapped in their control center, it's man versus machine in an all-too-real Wild West showdown.

A robot technician works on a glitch with the infamous gunslinger android in the movie *Westworld*.

Robots on Film continued

Even with their occasional menace and malfunctions, there are some movie robots you just can't help but love. Underneath their metallic shells and intertwined in their chips and circuitry is something almost resembling a human heart. These are robots you could take home to meet your mom.

Warm and Fuzzy

In the 1986 comedy *Short Circuit*, a lightning strike brings a military robot called SAINT Number 5 to life. In a confused state, the wobbly Number 5 wanders from its laboratory home and into the life of a free-spirited woman who proceeds to introduce her new friend to all sorts of cultural curiosities. As he learns more, he becomes more human—even calling himself Johnny 5—and tries to elude military personnel who want to destroy him. Though there were no shortage of actors in this film, most of the production budget was spent on nice guy Number 5.

Johnny 5, the fun-loving robot from the movie *Short Circuit*

Woody Allen impersonates a servant robot, also known as a Domesticon, in the hilarious film, *Sleeper.*

COMIC RELIEF

IN WOODY ALLEN'S uproarious futuristic comedy *Sleeper,* a nerdy health food store owner named Miles Monroe wakes up in the twenty-second century to find that the world he once knew (during the 1970s) has changed drastically. As he struggles with his new surroundings, Miles encounters several different kinds of robots, including house servants and Jewish tailors. When he is given a robotic dog named Rags as a pet, Miles asks: "Is he housebroken, or is he going to leave batteries all over the floor?"

Despite his negative attitude, it's hard not to like Marvin in *The Hitchhiker's Guide to the Galaxy.*

ROBOT DEPRESSION

FIRST APPEARING in 1978 as a character in Douglas Adams' radio series, *The Hitchhiker's Guide to the Galaxy*, Marvin the Paranoid Android made it to the big screen in 2005. Despite his name, Marvin isn't paranoid (which means very nervous)—he's just very, very, depressed. He's so intelligent that nothing he can do occupies enough of his brainpower, and he despises his shipmates on the *Heart of Gold.* This amazing ship carries Marvin (and the other characters) through deep space, but none of it is very interesting to this overly-intelligent robot. Marvin especially dislikes the automatic doors on the ship, which are always cheerful.

A distant relative of Johnny 5 is a rusty little robot named WALL-E from the 2008 animated movie of the same name. This tiny 'bot's unenviable task is cleaning up a horribly polluted planet Earth of the future. Since the remaining inhabitants of Earth have been relocated to a giant spaceship while the planet is cleaned up, WALL-E finds himself all alone. When a shiny robot named EVE unexpectedly arrives one day from the mother ship, WALL-E has one shot to experience true love and maybe even save humanity in the process.

WALL-E's name is actually an acronym for Waste Allocation Load Lifter Earth-Class.

Robots on TV

Since the early days of television, robots have been popular characters on science fiction–themed series and adventure shows.

A Boy and His Robot

One of the most popular TV robots was the protective B9 from *Lost in Space.* This show, which began in 1965, was also set in the future. In it, the Robinson family, including 9-year-old Will, set out in their spaceship, the *Jupiter 2,* to settle on a space colony. Their plans fall apart though, when the evil Dr. Smith reprograms B9 to destroy *Jupiter 2*. But Dr. Smith's plan fails, and the general utility robot becomes young Will Robinson's pal. The rest of the series focused on the Robinsons' adventures as they tried to find a way home. A lot of people remember how B9 waved its arms and yelled, "Warning! Warning! Danger, Will Robinson!" whenever his human pal faced a potential hazard.

The cast of *Lost in Space* with its robot star front and center

The *Lost in Space* robot often found itself on the receiving end of insults courtesy of Dr. Smith. Among the putdowns were "mechanical misfit" and "bumbling bag of bolts."

NOT A ROLE MODEL

DELINQUENT ROBOT Bender Bending Rodríguez, from the TV series *Futurama*, drinks, smokes, and gambles. He has to drink constantly because alcohol powers his fuel cells. He also constantly steals. Bender's secret passions are cooking and folksinging, and when he gets scared, bricks get ejected from his backside. His mother was a mechanical robot manufacturing arm, and his grandmother was a bulldozer.

Futurama's Bender is a robot with a real attitude and some interesting hobbies.

Lt. Commander Data is the smartest crewmember aboard the USS *Enterprise*.

A Robot Maid

The Jetsons, an animated series from 1960s, was about a family living in a space-age future. The show featured "aerocars" and lots of wacky futuristic inventions. And, of course, the show's creators couldn't imagine a future without robots that saved people from work–like Rosie the Robot. Rosie was the Jetsons' maid and kept their house spotless!

Rosie the Robot, the no-nonsense maid from *The Jetsons*

ROBOT DATA

Lt. Commander Data (played by Brent Spiner) in *Star Trek: The Next Generation* is super-smart, but like Mr. Spock before him, Data can't quite understand what the deal is with all these human emotions! Unlike Spock, who has no desire for such unnecessary baggage, Data can install an emotion chip so that he too can experience this strictly human behavior.

The Robot Hall of Fame

What better to pay tribute to some amazing robots than making a new hall of fame to honor them? In spring 2003, that is just what the School of Computer Science at Carnegie Mellon University in Pittsburgh, Pennsylvania, did. By November of the same year, the new Robot Hall of Fame had its first four members. Since then, a jury of scholars, researchers, writers, robot designers and builders, and businesspeople select real and fictional robots for recognition and induction into the Robot Hall of Fame.

Carnegie Science Center's RoboWorld in Pittsburgh, the largest permanent robotics exhibit in the United States, is home to the Robot Hall of Fame.

STAR WARS STAR BOTS

R2-D2, AN ASTROMETRIC DROID, helps pilots fly fighter spacecraft. After saving Queen Amidala's spaceship, R2 (or Artoo) helps Quigon Jinn and Obi-Wan Kenobi in *The Phantom Menace*. Later, he becomes the fast friend of C-3PO, a junked droid rebuilt from space parts by 9-year-old Anakin Skywalker to help his mother around the house. R2's other owners are Captain Antilles, Princess Leia, and Luke Skywalker.

A protocol droid capable of speaking millions of languages, the very proper C-3PO sounds British. The droid meets and befriends R2-D2 and the two robots form a buddy team, lending comedic relief to all the action of the *Star Wars* series. Like R2, C-3PO has appeared in all six *Star Wars* movies. C-P30 followed his friend into the Robot Hall of Fame in 2004.

Robot buddies C-3PO and R2-D2 have appeared in all six *Star Wars* films, making them, arguably, the most famous robots in the world.

The Sojourner rover successfully explored the surface of Mars back in 1997.

Science and Science Fiction

Because movie and TV robots often inspire robotics engineers to try to build robots that are just as complex, the founders of the Hall of Fame decided to select robots from two categories: Robots from Science and Robots from Science Fiction.

The Hall of Fame describes Robots from Science as "real robots that have served useful or potentially useful functions and demonstrated unique skills in accomplishing the purpose for which they were created. These may also be robots created primarily to entertain, as long as they function autonomously." Robots from Science Fiction are "fictional robots that have inspired us to create real robots that are productive, helpful, and entertaining. These robots have achieved worldwide fame as fictional characters and have helped form our opinions about the functions and values of real robots."

The First Four

The first two science robots inducted into the Hall of Fame were Unimate and Sojourner. Unimate, created by George Devol in the 1950s, was the first industrial robot. The NASA robot rover Sojourner explored Mars in 1997 as part of the Mars Pathfinder mission.

HAL 9000, the computer system from the 1968 movie *2001: A Space Odyssey* was one of the two first science-fiction robots to become a Hall of Famer. R2-D2, the little droid from the Star Wars series, was the other. R2-D2 came equipped with many gadgets and tools, including a gripping tool, a taser weapon, a computer interface, and a periscope. His gadgets often came in handy when he and his owners needed to escape some danger.

Hall of Fame Candidates

Lots of robots are being considered for membership in the Hall of Fame, including Swiss inventor Henri Maillardet's Automaton; Optimus Prime (from *Transformers*); Heathkit HERO (personal robots that educated their owners about robots); movie hero WALL-E; GIR (from *Invader Zim*); Mega Man (from the video game); AQUA (which explores underwater habitats; Huey, Dewey, and Louie (from the movie *Silent Running*); and Roomba (a self-operating vacuum cleaner).

BOSSY AND CRAZY

THOUGH HAL 9000 has a place in the Robot Hall of Fame, some critics would argue that it's not really a robot as much as it's a powerful computer. In the movie, HAL watches all aspects of the *Discovery* spaceship with its network of cameras and controls all of its operations. This ultimate artificial being is built to be perfect but runs into trouble with specific programmed instructions. The conflict between its hard-wired "perfect" instructions and the programmed instructions causes HAL to have a nervous breakdown, destroy the crew, and endanger the mission.

Even though HAL, the computer that ran things on the *Discovery* spaceship in the movie *2001: A Space Odyssey*, may not fit the standard idea of a robot, he achieved Robot Hall of Fame status.

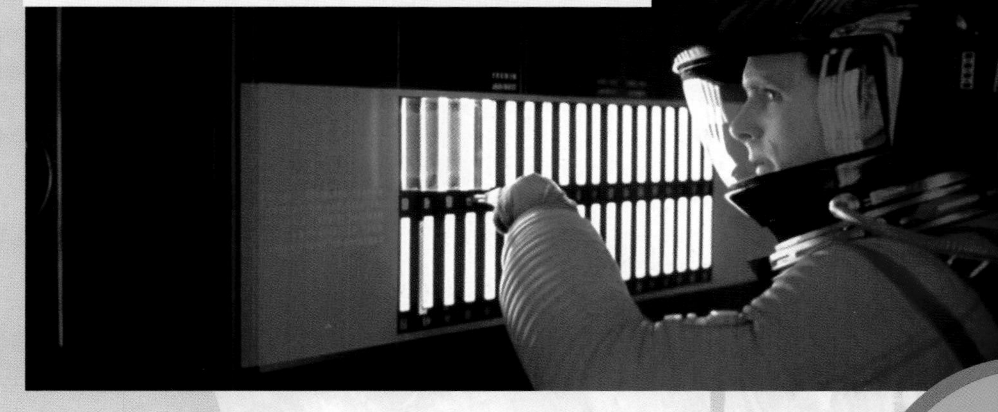

The Robot Hall of Fame

Since its beginning in 2003, the Robot Hall of Fame has added members every year or two. Each time it inducts new robots, it holds a ceremony to announce the inductees. Anthony Daniels, the actor who played the robot C-3PO in the Star Wars movies, has been the emcee for two of the ceremonies.

2004 INDUCTEES

In 2004, there were three science-fiction robots and two science robots inducted.

Science Fiction

- **Astro Boy.** Astro Boy was a powerful robot from the animated series *Astro Boy*, which first appeared in Japan in 1963.
- **C-3PO.** C-3PO was a protocol robot from *Star Wars*.
- **Robby the Robot.** Robby the Robot, actually a 7-foot, 2-inch tall robot suit worn by an actor, first appeared in 1956 in the movie *Forbidden Planet*. Robby went on to appear in dozens of other movies and TV shows.

Science

- **ASIMO.** ASIMO is a humanoid robot created by Honda Corporation.
- **Shakey the Robot.** Shakey, developed by the artificial intelligence laboratory of SRI International, was the first mobile robot that could reason about its own actions. For example, when an operator, at the computer station that Shakey was connected to, typed in the command, "push the block off the platform," Shakey would look around and identify a platform that had a block on it. It would then locate a ramp, roll up it, and then push the block off the platform.

2006 INDUCTEES

WHEN THE HALL OF FAME decided on its next batch of inductees, fictional robots again outnumbered the real ones.

Science Fiction

- **Maria.** Maria was the robot woman from the 1927 movie *Metropolis.*
- **Gort.** Gort was an 8-foot-tall humanoid robot from the 1951 movie *The Day the Earth Stood Still,* which was remade in 2008. Gort doesn't do much except stand silently in front of his spaceship. He never speaks, but he comes equipped with a laser weapon that shoots beams from his visor, which he uses to vaporize weapons and obstacles. He is on Earth as part of an interstellar police force whose purpose is to preserve peace all over the universe.
- **David.** David was the android-kid from *A.I.: Artificial Intelligence* (2001).

Science

- **AIBO.** AIBO was Sony's cyber-dog.
- **SCARA** (Selective Compliance Assembly Robot Arm). SCARA is a class of robot arms developed in the 1970s and 1980s that can pick and place. They have many industrial uses.

Robby the Robot's space age design inspired many imitators.

2007–08 INDUCTEES

THE CEREMONY for the 2007–2008 was the first to see more science robots inducted than science-fiction ones.

Science Fiction

- **Lieutenant Commander Data.** Data was the android from *Star Trek: The Next Generation.*

Science

- **LEGO Mindstorms NXT.** Mindstorms are kits that let you build your own robot.
- **NAVLAB.** NAVLAB is a minivan that drives itself. Its greatest feat was its "no hands across America" journey from Pittsburgh, Pennsylvania, to San Diego, California. The human driver barely ever needed to touch the steering wheel during the cross-country drive.
- **Raibert Hopper.** The Raibert Hopper, built by Marc Raibert, was a one-legged robot built to test balance. It successfully bounced around on its single leg.

In the future, ASIMO may be used to assist people with disabilities or the elderly.

DREAM: IMPOSSIBLE

ASIMO gets its name from science-fiction writer Isaac Asimov, and the name stands for Advanced Step in Innovative Mobility. This 4-foot, 3-inch-tall robot can run, walk on uneven slopes and surfaces, turn smoothly, climb stairs, grasp objects . . . and conduct a symphony orchestra. In April 2008, ASIMO conducted the Detroit Symphony Orchestra playing "The Impossible Dream." ASIMO can also understand and respond to simple voice commands and can recognize and remember individual faces. Using its camera eyes, ASIMO can map its environment and avoid bumping into objects or walls.

In the film *AI: Artificial Intelligence*, Haley Joel Osment plays a robot named David that has been programmed with the ability to love its owners like a child loves a parent.

Robots of the Future

In the not-so-distant future, robot fans can look forward to a new generation of intelligent multitasking machines moving out of the factory and into the home and general workplace. In a matter of a few years, the problem of movement—making robots that can walk elegantly and easily—will be solved, and the science of programming them to learn from their mistakes will improve. Advances in computer science will enable the average robot to interpret people's intentions and predict actions. The decades ahead will see humanoid robots that can help the disabled and perform tasks such as handling luggage and giving tours. Researchers are already working on robots, like HUBO Lab's Albert HUBO, that can walk, talk, and recognize individuals: all necessary steps to developing the robots of the future.

Albert HUBO, designed to look like its namesake, famous physicist Albert Einstein, stands four and a half feet tall and weighs 126 pounds. His programmers have taught him human gestures, like waving, which will be an important feature of any household robot of the future.

BRAINY ROBOTS

REVERB (short for Reverse Engineering the VERtebrate Brain") is a British project to create an intelligent robot that can interpret and react to its environment. The developers hope to complete the project in 2010 and hope that it will help scientists to better understand how the brains of animals work.

REVERB is an attempt to create animal-like artificial intelligence.

Site of the Stanford AI Lab (SAIL) — Stanford, CA

Site of McGill Centre for Intelligent Machines — Montreal, **CANADA**

Site of CalTech Robotic Group — Pasadena, CA

Site of Case Western Reserve Biologically Inspired Robots Laboratory — Cleveland, OH

Site of MIT Field and Space Robotics Laboratory — Cambridge, MA

NANOBOTS

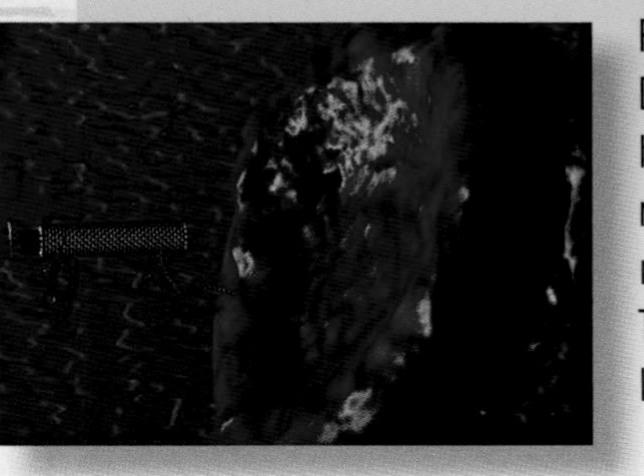

An artist's drawing of a nanobot in the bloodstream.

RESEARCHERS AT DARTMOUTH UNIVERSITY have built an inchworm-like robot so small you need a microscope just to see it. Two hundred of them could line up across a single plain M&M. The nanobot is longer than it is wide—but since it's only about as wide as a human hair, it's still the smallest untethered (not attached to anything) controllable microrobot ever. The robot crawls like a silicon inchworm, making tens of thousands of miniscule steps every second. It turns by putting a silicon "foot" out and pivoting like a motorcyclist skidding around a tight turn. With this innovative bending movement, it can move freely across a surface without the wires or rails that restricted the mobility of previously developed microrobots. Later versions might inspect and make repairs to integrated circuits, explore hazardous environments, or even be injected into humans to explore or manipulate cells or tissues.

The field of robotics continues to grow, with research performed around the world.

Site of Mobile Robotics Research Group

Edinburgh, **SCOTLAND**

Site of Dalle Molle Institute for Artificial Intelligence

Manno-Lugano, **SWITZERLAND**

Site of Automation Technology Laboratory at Helsinki University of Technology

Helsinki,

Site of University of Sheffield Robotics and Industrial Automation Department

Sheffield, **UK**

Site of computational Vision and Robotics Laboratory at the Institute of Computer Science

Crete, **GREECE**

DID YOU KNOW?

NASA's Wolfgang Fink is developing thinking robots to pair with dumber robot explorers. The thinking robot would orbit a planet while its less intelligent partner would roam the surface. What's the advantage to humans? Well, it's tough exploring the outer planets, even with rovers like the Mars landers, because it would take so long to send signals from Earth to a robot on a distant planet. Having an intelligent robot in orbit means it could coordinate rover missions without lengthy days. These robot orbiters could also react immediately to disasters. Eventually, NASA may develop explorer robots like Disney's Eve (from WALL-E), which would be completely autonomous.

Find Out More

WORDS TO KNOW

AIBO. Sony's programmable robotic canine toy, the first of its kind

amphibious. Capable of functioning on land or in water

android. Robot that looks and acts like a human

articulating. Able to make flexible movements

Artificial Intelligence (AI). Machines that can think on their own

autism. A childhood development disorder concerning limited communication skills and difficulty interacting with others

automaton. Mechanical figure that performs human tasks

bionic. Robotic part that mimics human muscles or a human organ or limb

Bluetooth. Wireless device linking people or robots to a computer or cell phone

Braille. A system of writing and printing that uses raised dots; used by blind and visually impaired people

component. Part of a larger machine, object, or project

coral. Hard and often multicolored skeletons of marine life

cyborg. Combination of human and machine

data. Information

defuse. To deactivate an explosive device

disabled. Deactivated (for machines), or those with limited movement or function (for people)

droid. Star Wars nickname for any robotic servant

drone. A remote-controlled air, sea, or land vehicle

foundries. Factories that specialize in pouring molten metal into molds

functioning. Working

gadget. A small hand-sized machine or device

geology. The study of a planet's soil and rocks for historical information

gyroscope. A stabilizing device, consisting of a wheel mounted on an axis, which aids in maintaining balance and direction

humanoid. Having human characteristics or form

infrared. A wavelength just outside those that are visibly recognizable, such as red light. These rays are often used for such things as television remote controls.

MIT. Massachusetts Institute of Technology, a university with a large division devoted to robotic development

microprocessor. A circuit that holds most of a computer's central processing functions on one chip.

multi-articulating. Able to have two or more flexible sections or extensions

multitasking. Able to perform two or more actions at the same time

nanotechnology. Science of using microscopic machines to perform tasks inside a machine or human

paralyzed. Unable to move

PC (Personal Computer). A computer designed for individual use

piston. A cylinder-shaped piece of machinery that moves in a back-and-forth motion caused by pressure from a fluid

programming. Process of entering instructions to a computer

Renaissance. A period in Europe between the fourteenth and seventeenth centuries known as a time of rebirth for the arts and learning

robotics. The science and/or technology associated with robots

Roomba. A self-operating vacuum cleaner

rover. Exploratory robot used in interplanetary or underwater missions

rotary joint. A joining point between two pieces that allows rotational movement

scaffold. A temporary platform used to hold workers and materials

sensor. Electronic device that can react to or gather data about a stimulus such as sound, light, or motion. Sensors can be used to send information to computers.

software. The programs (or instructions) that run a computer

sophisticated. Complicated; not simple

squadron. An air force term, used to describe a unit of two or more airplanes

surveillance. The act of watching or observing a person or situation

toxins. Poisonous substances

ultrasound. A very high frequency of sound that humans cannot hear

Universal Serial Bus (USB). Simple cable used to connect printers, scanners, and other outside devices to a computer

Unmanned Air Vehicle (UAV). A remote-controlled aircraft

untethered. Free of any attachment or connection

Wi-Fi. A wireless computer networking device

WEB SITES TO VISIT

Battle Kits

www.battlekits.com

Seller of products for robotic education and competition, including different kits for different weight classes, from featherweight (30 pounds) to heavyweight (220 pounds to 340 pounds).

Doctor's Gadgets

http://www.doctorsgadgets.com/building-the-bionic-man-from-eye-to-anus.html

This site provides news on all types of medical breakthroughs in bionics, from eye, ear, even tongue and brain to heart, kidneys, liver, legs, and more!

Hitec Robotics

www.hitecrcd.com and www.hitecrobotics.com

Hitec is dedicated to providing quality, affordable radio-controlled hobby products everyone can enjoy. One of the company's latest, and most successful, kits is the RoboNova-1.

Robot Galaxy

www.robotgalaxy.com

People can assemble, program, and activate their own design of robot at three stores in and around New York City. Robot Galaxy offers a comic book featuring robot heroes the Brotherhood that kids can read about and build from kits. Each Robot Galaxy features a bridge, flight deck, and mission control.

Robot Magazine

www.botmag.com

Robot magazine offers news and views on technological advances in robotic as well as advice to the robot hobby enthusiast and collector.

The Robot Shop

http://www.robotshop.us

www.robotshop.ca

This online store sells state-of-the-art robot kits, as well as personal and professional robots and repair.

Robot Village

www.robotvillage.com

Home to the Robot of the Month Club for kids from eight to ten, this place offers robot kits and online classes.

BOOKS TO READ

I, Robot by Isaac Asimov

If I Had a Robot by Dan Yaccarino

If I Had a Robot Dog by Andrea Baruffi

Marveltown by Bruce McCall

123 Robotics Experiments for the Evil Genius by Myke Predco

Ricky Ricotta's Mighty Robot Series by Dav Pilkey, illustrated by Martin Ontiveros

Robot Dreams by Sara Varon

Robots, Androids and Animatrons: 12 Incredible Projects You Can Build by John Iovine

Robot Companions: MentorBots and Beyond by E. Oliver Severin

The Robot Builder's Bonanza by Gordon McComb and Myke Predko

The Robot Series (also by *Isaac Asimov*)

The Robot Builders Sourcebook: Over 2,500 Sources for Robot Parts by Gordon McComb, publisher

The Unofficial Lego MINDSTORMS NXT Inventor's Guide

Index

Credits

Abbreviations Used

AP = Associated Press; iSP = *iStockphotos.com*; JI = *Jupiter Images*; LoC = *Library of Congress*; NASA = *National Aeronautics and Space Administration*; NOAA = *National Oceanic and Atmospheric Administration*; PF = *Photofest*; PR = *Photo Researchers*; SS = *Shutterstock*; Wi = *Wikimedia*

l = left, *r* = right, *t* = top, *b* = bottom; *m* = middle

3 SS/Argus **6** SS/Diana Rich **6**l SS/Baloncici **6**r SS/MAMEL **7**t PF **7**b PF **8–9** PF **10**l PR/Makoto Iwafuji **10**r PR/Gustoimages **11**tr PR/Pasquale Sorrentino **11**bl PR/Victor Habbick Visions **12**l PR/Edward Kinsman **12**r PR/Mauro Fermariello **13** AP/Itsuo Inouye **14**l PF **14**r PF **15** Wi/Erik Möller **16** PF **17**tl Popular Science **17**tr Science and Invention Magazine **17**bl PF **17**br Wi/George C. Devol **18–19** SS/Chris Harvey **20**bl Wi/Manfred Werner **20**tr PR/Massimo Brega **21** Wi/Jared C. Benedict **22**tr Wi/RobotWorx **22**bl SS/Baloncici **23**tr SS/Baloncici **23m**r SS/SFC **23**bl SS/Jonathan Heger **24**tr © 2006, Intuitive Surgical, Inc. **24**br © 2006, Intuitive Surgical, Inc. **24**bl © 2006, Intuitive Surgical, Inc. **25** Kokoro Company Ltd. **26** r U.S. Army **26**l PR/Victor Habbick Vision **27**tl U.S. Army **27**br U.S. Marine Corp **27**bl U.S. Marine Corp **28** AP/Jill Connelly **29**t U.S. Air Force **29**b SS/Paul Drabot **30**l United Service Associates, Inc. **30**r U.S. Air Force **31**t SS/Stavchansky Yakov **31**b SS/Stavchansky Yakov **32**l NASA **32**r NASA **33** NASA **34–35** Cornell University **34**l NASA **34**r NASA **35**t NASA **35**b NASA **36**l JI **36**r NASA **37** NASA **38**tr PR/David Vaughan **38**b NOAA **39**tl NOAA **39**b PR/A. Gragera, Latin Stock **40**tr Wi/Denoir **40**tl Wi/Shawnc **40**bl iSP/Fanelli Photography **41**tl PR/Hank Morgan **41**br Wi/BradBeattie **42** NEC System Technologies **43** PR/Marcello Bertinetti **44**l Wi **44** r Wi **45**t AP/STR **45**b Wi/Aaron Biggs **46**l PR/Mauro Fermariello **46**r IZI Robotics **47**t NEC System Technologies **47**b Wi/Bevan Weir **48** PR/David Grossman **49**l Wi/Dfrg.msc **49**r Lynxmotion, Inc. **50–51**c AP/JOERG SARBACH **50**bl Carnegie Mellon University **50**br AP/Itsuo Inouye **51**br AP/Sang Tan **52–53** SS/Juanjo Tugores **54**l PF **54**r PF **55**tl PF **55**b PF **56**l PF **56**r PF **57**l SS/Andreas Meyer **57**r PF **58**l PF **58**r PF **59**l PF **59**r PF **60**l PF **60**r PF **61**tl PF **61**bl PF **61**r PF **62** PF **63**tl NASA **63**br PF **64** PF **65**l Wi/Xavier Caballe **65**br PF **66**tr WI **66**bl NASA **67** Rutgers University

Backgrounds SS/argus

Cover SS/argus